ALGEBRA EXAMPLES

POLYNOMIAL FACTORIZATIONS 2

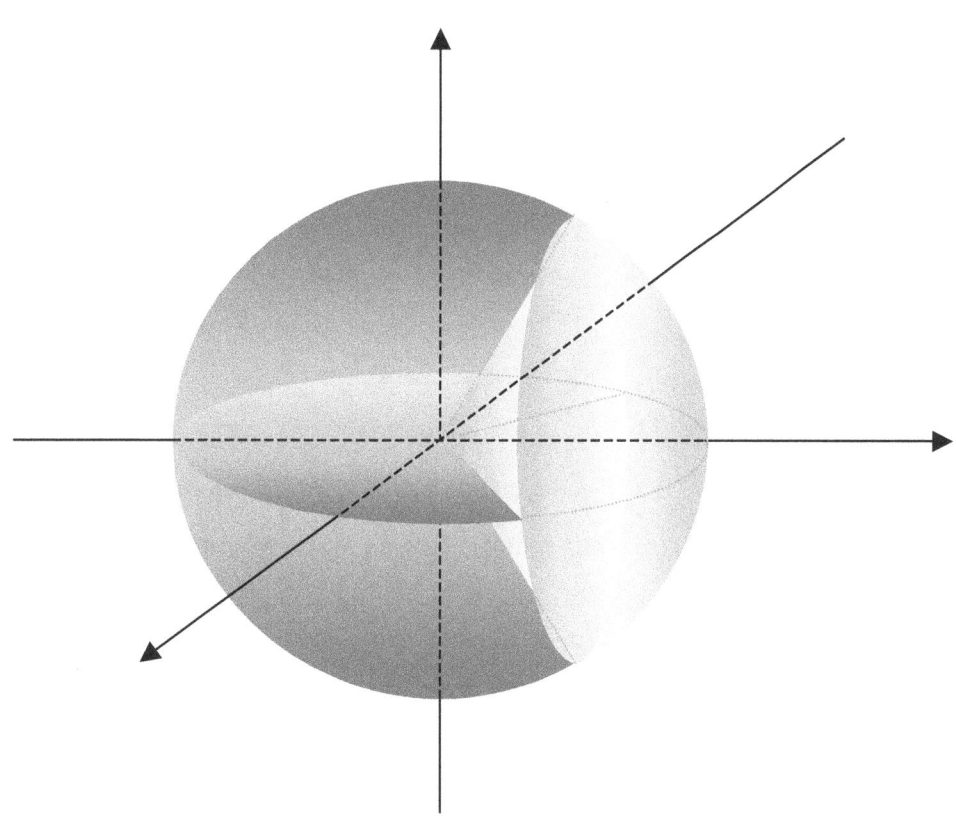

Seong R. KIM

Dear students:

Students need the best teacher, so you need examples, because examples are the best teacher. All the examples here are fully worked, and explain **how** the basic and essential tools in math are made, together with **what** they are, **how** they work, and **how** to work with them. Such tools include numbers, formulas, identities, equations, laws, etc.

Examples here begin with easy ones, of course. Covering every meter and yard properly, we can cover thousands of miles and kilometers. And it is particularly the case in math.

Of those examples therefore, some might even look too easy for you. It's not that easy though, to come up with those examples. Anyways, the bigger and the taller the tree, the deeper and the stronger the root.

Doing math, we work with ideas and run ideas, because every thing in math is an idea. A number is an idea, for instance, and the same is true for a line or circle, too. And putting ideas together, we build another, which becomes the base or an element of another, and each is connected. And that's the way your math grows. So you get to build a circuit, and sometimes, need to fill the gap or repair the circuit so that you get the sense of it.

So your calculation runs properly, and you get the problem solved.

The examples have been made and arranged so that they get tougher (or sometimes easier for some reason) as you proceed with them. In particular, similar examples with some variations are strategically repeated so that you can get the ideas or the tools tricky or complicated, and can get them mastered.

This book is however, nothing but a bunch of examples until you get it powered.　How then, to get it powered, and make it run and work for you?

Just read it, and then, do each example in writing. And it is important to note that you do it in **your** writing. Just watching someone doing it, you just only feel that you can do it. If you do it, you can do it, but if you don't, we can hardly. It's a cliché, of course, but is always true that knowing is one thing and doing is another.

I've been helping students grow, take care of, and run their own math. The area covers algebra and geometry for high school or college students, and is especially for equations (for unknowns or curves), functions, and their graphs, which are the basic elements in calculus, which's been the core of my interest from my early age in high school.

Of my students, some are quite poor in math, and thus, are afraid of or hate math, some require special education because of exceptional intelligence, some are smart enough, some are naïve and diligent, some are clever but lazy, and most behave in general. All the students are badly after though, one thing in common: a strong and secure math skill. It is of course, the prime objective of my work, and I'm always happy to and eager to help them achieve it. The problem was however, that many of them wanted it to be purchased. And the question is, can we buy it?

We can buy the means, of course. And a solid math skill is feasible, too. We know however, we can't buy love, and the same is true for the math skill, too. It's not what we can buy or sell, and not what we can give or take. It is however, what we can grow, and need to grow. Your math grows as much as you grow and take care of it. So does mine.

What math then, do students most often do or use in high schools or colleges?

It is algebra and geometry. What algebra though?

Elementary algebra, of course
Doing the algebra, we work with numbers (many in kinds), constants, variables, ratios, rates, expressions, equations, inequalities, functions, identities, formulas, laws, etc., together with signs and symbols. And if we want to do algebra properly, we want to know their natures and how they mingle with each other.

So studying math ideas or tools, you want to know **what** they are, **how** they work, and **how** to work with them or **what** to do with them.   What then, about the geometry?

Basically, the geometry has much to do with shapes, positions, and angles. The shapes begin with triangles and circles, and move on to rectangles, squares, parallelograms or rhombuses, trapezoids, tetragons, other polygons, polyhedrons, etc.

Doing the geometry, too, though, we need to do the algebra stated above. So it is analytic geometry, often called coordinate geometry, too. And doing it, we can specify positions using coordinates. So in the geometry, basically, we work with graphs. Putting a math idea in a graph, we can not only effectively think about it but actually see it, too, and therefore, can efficiently work with it. What idea then, is it?

The idea begins with a point, line, parabola, circle, ellipse, and hyperbola, called a conic section or basic curve, and then, moves on to other curves, planes, surfaces, volumes, and other objects in various dimensional spaces, together with vectors.

And using an angle, we can specify an amount of turn or change in direction.

So learning, using, or applying those ideas or math tools, we get to solve problems.

And this book can help. It can help learn them, and use them so that you can navigate to find solutions to problems. And in particular, it can help come up with answers to those **what**s and **how**s stated above. So it can help you grow and run your own math, and thus, can help achieve your solid math skill.

It is however, not a magic book giving you a math skill of high caliber overnight. And it can have many mistakes, too. There is no magic, and math is full of facts and ideas. And it is after all, not me and not your teacher but you who put together some of those facts and ideas, and understand it. Putting facts and ideas together, understanding it, and taking care of what you have learned, you grow your math. And this book can help.

This is a book of examples designed to help you grow your math, and assumes that you are a real beginner. This book requires though, time and effort, the amount of which need to be substantial, too, but will be worth it. That's because you want a substantial achievement, and will get it. And probably, you will get to see this book helping you get there much faster than expected. And then, you will get to see the way math runs.

In math, everything is an idea. So is a problem. And solving it, we put it many different ways. For instance, while expanding or reducing it, or modifying or converting it, we keep searching for the solution, approaching the solution, and eventually, can get there. So don't look for the solution outside the problem. The solution is inside the problem if the problem is properly made.

If it is not, no solution is the solution. And in fact, it is often the case a problem itself is the solution. We can put a problem in many different ways, and eventually, can end up with the solution.   How come then, is the solution no other than the problem?

For instance, the solution to $3232 \div 101$ is 32.   And we can put it this way:

$$3232 \div 101 = \frac{3232}{101} = \frac{32 \times 101}{101} = \frac{32}{1} = 32 \Rightarrow 3232 \div 101 = 32.$$

And we can get this, too: $32 \Rightarrow 3232 \div 101.$ How?

$$32 = \frac{32}{1} = \frac{32 \times 101}{101} = \frac{3232}{101} = 3232/101 = 3232 \div 101.   \text{Too easy?}$$

For another instance, the solution to $ax^2 + bx + c = 0$ is: $x = \frac{-b \pm \sqrt{b^2 - 4ac}}{2a}$, which is called the quadratic formula.   How come then, is the solution no other than the problem?

We can put it this way:

$$x = \frac{-b \pm \sqrt{b^2 - 4ac}}{2a} \Rightarrow 2ax = -b \pm \sqrt{b^2 - 4ac} \Rightarrow 2ax + b = \pm\sqrt{b^2 - 4ac}$$

$$\Rightarrow (2ax + b)^2 = b^2 - 4ac \Rightarrow 4a^2x^2 + 4abx + b^2 = b^2 - 4ac$$

$$\Rightarrow 4a^2x^2 + 4abx = -4ac \Rightarrow ax^2 + bx = -c \Rightarrow ax^2 + bx + c = 0.$$

And we can get this, too: $ax^2 + bx + c = 0 \Rightarrow x = \frac{-b \pm \sqrt{b^2 - 4ac}}{2a}.$ How?

$$ax^2 + bx + c = a(x^2 + \tfrac{b}{a}x) + c = a(x^2 + \tfrac{b}{a}x + \tfrac{b^2}{4a^2} - \tfrac{b^2}{4a^2}) + c = a(x^2 + \tfrac{b}{a}x + \tfrac{b^2}{4a^2}) - \tfrac{b^2}{4a} + c$$

$$= a(x + \tfrac{b}{2a})^2 - \tfrac{b^2 - 4ac}{4a} = 0 \Rightarrow a(x + \tfrac{b}{2a})^2 = \tfrac{b^2 - 4ac}{4a} \Rightarrow (x + \tfrac{b}{2a})^2 = \tfrac{b^2 - 4ac}{4a^2} \Rightarrow x + \tfrac{b}{2a} = \pm\sqrt{\tfrac{b^2 - 4ac}{4a^2}}$$

$$\Rightarrow x = -\tfrac{b}{2a} \pm \tfrac{\sqrt{b^2 - 4ac}}{2a} = \tfrac{-b \pm \sqrt{b^2 - 4ac}}{2a} \Rightarrow x = \tfrac{-b \pm \sqrt{b^2 - 4ac}}{2a}.$$

And we call the set of processes above, algebra.

So if a problem is well defined, that is, if it makes sense, we should be able to get it solved the way below:

A problem $\Rightarrow$ **…** $\Rightarrow$ **…** $\Rightarrow$ **the solution**, and thus: **the problem** $\Rightarrow$ **the solution**.

So solving a problem, we put it many different ways so that we can get to the solution.

And that's the way, math runs.

May your math run very well.

Seong R. Kim

B.S. Math. Michigan Tech. Univ.   M.S. Math. Rensselaer Polytechnic Institute

Notes:

This book is about factorizations of polynomials.
Factorization is often called factoring, and is a math operation.

Factorizing a polynomial, that is, factoring a polynomial, we get to at the same time do all the four arithmetic operations: additions, subtractions, multiplications, and divisions. So we get to do a lot of mental math doing factorizations. Doing them, we don't just do mental math but we also need to apply rules or laws between the operations above, and the operations are done on not just numbers but expressions, too. So we need to do a lot of algebra.

Factorizing a polynomial, we break it into pieces or parts, and put them together to get another expression, which is a product of expressions as $2x(x + 1)(x - y + 3)$. And doing algebra, we often need to do such math operations to change expressions. Why changing them?

That's simply because we want to get to the solutions.
What actually, that is, physically, connects problems to solutions is algebra.
With algebra skill, together with your creativity, you can actually solve problems.

And most expressions are made of polynomials. So if you want to build strong skill in algebra, you want to be good at factorizing polynomials as well as integers.

And all the basics and many ideas on polynomial factorization are covered in two books as well as in one book. And the two books are as follows:

ALGEBRA EXAMPLES POLYNOMIAL FACTORIZATIONS 1

ALGEBRA EXAMPLES POLYNOMIAL FACTORIZATIONS 2

And the two books are combined into one book, which is below:
ALGEBRA EXAMPLES POLYNOMIAL FACTORIZATIONS

So either way, the books explain how to manipulate polynomials, that is, how to change or alter, convert, or modify expressions so that you can come up with the ones you need. The ones are solutions, of course. And that's what polynomial factorizations are about.

The books do not just explain. But they help you, too, follow steps to the solutions so that you can see how you can change expressions, and see how calculations can flow.

With strong foundation of algebra, you can do a lot, and of course, can do problems very well, too. And following steps in factorization processes, that is, changing expressions or taking alternatives shown in the books, you can grow much power in algebra.

So the books will get you not only polynomial factorizations but enhancement of your algebra, too. You will thus, soon be able to change or alter, convert, or modify math expressions so that you can get to the solutions fast.

Contents

In POLYNOMIAL FACTORIZATIONS 2

The Preview of the Contents

In POLYNOMIAL FACTORIZATIONS 1

$$(x + y)^2 = x^2 + 2xy + y^2. \qquad (x + y)^3 = x^3 + 3x^2y + 3xy^2 + y^3.$$

$$(x + y)(x - y) = x^2 - y^2. \qquad (x + y)(x^2 - xy + y^2) = x^3 + y^3.$$

$$(x^2 + xy + y^2)(x^2 - xy + y^2) = x^4 + x^2y^2 + y^4.$$

$$(x + a)(x + b) = x^2 + (a + b)x + ab. \qquad (ax + b)(cx + d) = acx^2 + (ad + bc)x + bd.$$

$$(x + a)(x + b)(x + c) = x^3 + (a + b + c)x^2 + (ac + bc + ca)x + abc.$$

$$(a + b + c)^2 = a^2 + b^2 + c^2 + 2(ab + bc + ca).$$

$$(a + b + c)(a^2 + b^2 + c^2 - ab - bc - ca) = a^3 + b^3 + c^3 - 3abc.$$

Suppose both a and $b \neq 0$, and both m and n are integers. Then, we get:

0. $a^m a^n = a^{m+n}$ **1.** $a^m / a^n = \dfrac{a^m}{a^n} = a^{m-n}$ **2.** $(a^m)^n = a^{mn}$

3. $(ab)^n = a^n b^n$ **4.** $(a/b)^n = \left(\dfrac{a}{b}\right)^n = a^n / b^n = \dfrac{a^n}{b^n}$

Suppose both a and $b > 0$, and m and n both are integers nonzero. Then, we get:

0.1. $a^{\frac{1}{n}} b^{\frac{1}{n}} = (ab)^{\frac{1}{n}}.$ **1.1.** $\dfrac{a^{\frac{1}{n}}}{b^{\frac{1}{n}}} = \left(\dfrac{a}{b}\right)^{\frac{1}{n}}.$ **2.1.** $(a^{\frac{1}{n}})^m = (a^m)^{\frac{1}{n}}.$

3.1. $(a^{\frac{1}{n}})^{\frac{1}{m}} = a^{\frac{1}{mn}} = (a^{\frac{1}{m}})^{\frac{1}{n}}.$ **3.2.** $(a^{mp})^{\frac{1}{np}} = (a^m)^{\frac{1}{n}}$, where p is a nonzero integer.

1. Suppose M, N, and $b > 0$, but $b \neq 1$, and we have: $A = \log_b M$, and $B = \log_b N$. Then, we get: $A - B = \log_b M - \log_b N = \log_b \frac{M}{N}$.

2. Suppose that M and $b > 0$, but $b \neq 1$, and that we have: $E = \log_b M$. Then, we get: $PE = P \log_b M = \log_b M^P$.

3. Suppose that a, b, C, and $D > 0$, but a and $b \neq 1$, and that we have: $\log_a C = \log_b D$. Then, we get: $\log_a C = \log_b D = \log_{ab} CD$.

4. Suppose that a, b, C, and $D > 0$, but a and $b \neq 1$, and that we have: $\log_a C = \log_b D$. Then, we get: $\log_a C = \log_b D = \log_{\frac{a}{b}} \frac{C}{D} = \log_{\frac{b}{a}} \frac{D}{C}$.

5. $\log_b b = 1$, and $\log_b 1 = 0$. **6.** $\log_b A = \dfrac{\log_c A}{\log_c b}$.

7. $\log_b A = \dfrac{1}{\log_A b}$.

Note:

The drawings or graphs in this book are not exact, and are approximate or conceptual ones.

$\in$	"$a \in B$" means that a belongs to B. "p, q, and $r \in W$" means that p, q, and r belong to W.
$\Rightarrow$	"$A \Rightarrow B$." means that A implies B.
$\equiv$	$A \equiv B$ means that A and B are identical to each other.
$\neq$	$A \neq B$ means that A is not equal to B.
$\lvert A \rvert$	The magnitude of A. For instance, $\lvert -1 \rvert = \lvert 1 \rvert = 1$.
$\therefore$	Therefore
$\Leftrightarrow$	"$A \Leftrightarrow B$" means "If A then B." and "If B then A." We can read $A \Leftrightarrow B$ as "A if and only if B." In such a case, we can say that $A = B$.
Δx and Δy	Suppose that (x_1, y_1) and (x_2, y_2) are two points in the x-y plane. Then, we get either of the two below. $\Delta x = x_2 - x_1$, and $\Delta y = y_2 - y_1$. $\Delta x = x_1 - x_2$, and $\Delta y = y_1 - y_2$.

Distance Formula

Suppose that d is the distance between two points (x_1, y_1) and (x_2, y_2) in the x-y plane. Then, we get: $d^2 = (\Delta x)^2 + (\Delta y)^2$.

Examples B

Why polynomial factorizations?

They are for your algebra skill. So knowing them very well and doing them very well, you can do very well manipulate, that is, change, convert, or modify math expressions.

So when studying this material, too, pay particular attention to the processes math expressions can be changed or modified through. Then, you can see better how to manipulate, that is, change or modify math expressions.

- It is in fact, your algebra that can actually connect the problems to the solutions.

While learning polynomial factorizations, you can improve your skill of algebra a lot. And those factorizations are great tools, and can be powerful if you know them very well, and do practices on them often enough.

Cliché but always true: Knowing is one thing and doing it is another.

So make your own examples, too, and do them yourselves.

Then, you can do better your algebra as well as see better the principles working behind the tools called polynomial factorizations.

Factorize the polynomials below.

0. $6x^2 + 23x + 20$

1. $28x^2 + 71x + 18$

Suggestions or Solutions
To the Problem in the Example 0

We have: $6x^2 + 23x + 20$.

Factorizing it, we can get it the way below:

$$6x^2 + 23x + 20 = 2{\cdot}3x^2 + (2{\cdot}4 + 3{\cdot}5)x + 4{\cdot}5 = 2{\cdot}3x^2 + 2{\cdot}4x + 3{\cdot}5x + 4{\cdot}5$$

$$= 2x(3x + 4) + 5(3x + 4) = (3x + 4)(2x + 5).$$

If not quite sure of the idea behind the processes above, follow the steps below:

In this example, we have several things to discuss besides the solution.
And the things are about growing your algebra, which is the purpose of this book.
That is to say that they are for the enhancement of your algebra skill.

Now, setting first, $P = 6x^2 + 23x + 20$, we can say that the polynomial P is quadratic, and is in the form of $ux^2 + vx + w$.   So?

The polynomial $ux^2 + vx + w$ is similar to a particular polynomial, which is factorize to $(ax + b)(cx + d)$.

What then, is the particular polynomial?

It is: $acx^2 + (ad + bc)x + bd$.

Taking ac as u, $ad + bc$ as v, and bd as w, we get: $ux^2 + vx + w$, which P is in the form of.
So P can be said to be similar to $acx^2 + (ad + bc)x + bd$, factorize to $(ax + b)(cx + d)$.

Thus, setting: $ac = 6$, $ad + bc = 23$, and $bd = 20$, and then, finding a, b, c, and d, we can get the factorization of P.   How come?

First, setting: $ac = 6$, $ad + bc = 23$, and $bd = 20$, we get:

$acx^2 + (ad + bc)x + bd = 6x^2 + 23x + 20$, which is the very P.

Next, we know: $acx^2 + (ad + bc)x + bd = (ax + b)(cx + d)$, so finding a, b, c, and d, we can get the factorization of P.

That is, we can set: $6x^2 + 23x + 20 = (ax + b)(cx + d)$.

That is to say that solving the system of equations below, we can get it done:

$ac = 6$, $ad + bc = 23$, and $bd = 20$.   Are they all the equations we can make, though?

In other words, is there any other equation we can add to the system?

That seems to be it. Then, we've got a problem.   What problem then, is it?

The number of equations has to be the same as the number of unknowns.
We have four unknowns, yet the system has three equations only.
So the system lacks one equation, and thus, is not complete.
So we've got a situation here.   What else then, can we do?

We can set up a reasonable constraint, which is on the unknowns, of course, and then, make use of the constraint.   Doing math in fact, we often do so.

That is to say that we can set up a reasonable condition that the unknowns have to obey.

Satisfying the condition, along with the three equations, we can find the four unknowns. That is because an equation is in fact, a condition to be met.

What do we mean by a reasonable condition though?

We can restrict the unknowns to be integers. How?

We want $P = 6x^2 + 23x + 20$ to be in the form of $acx^2 + (ad + bc)x + bd$, which is factorized to $(ax + b)(cx + d)$.

So setting: $ac = 6$, $bd = 20$, and $ad + bc = 23$, we can put P into the form above.

And we know divisors of an integer are integers, a and c are divisors of ac, that is, 6, and b and d are divisors of bd, that is, 20.

So we can set up a condition that a and c are two integers the product of which is 6, and also, b and d are two integers multiplied to be 20.

So *if any*, we can find the four integers a, b, c, and d satisfying the equations below:

$ac = 6$, $bd = 20$, and $ad + bc = 23$.

That is to say that the system is now: $ac = 6$, $bd = 20$, $ad + bc = 23$, and a, b, c and d are integers. So the system has 4 equations. How come though, the system has 4 equations?

An equation is in fact, a condition to be satisfied.

And the system above can be said to have four conditions. And the four are as follows:

- $ac = 6$.
- $bd = 20$.
- $ad + bc = 23$.
- a, b, c and d are integers.

So the system is well defined. What do we mean by a system well defined, though?

If a problem or a system of equations is well defined, it is workable.

What do we mean by though, "*If any*" in the statement below?

If any, we can find the four integers *a*, *b*, *c*, and *d* satisfying the equations below: *ac* = 6, *bd* = 20, and *ad* + *bc* = 23.

It means that it can be the case where there do not exist four integers that can be *a*, *b*, *c*, and *d*. Then, we cannot get the factorization of *P* within the scope of integers.
What then, do we do?

We want to use the quadratic formula, which is: $x = \frac{-b \pm \sqrt{b^2 - 4ac}}{2a}$, which is the solution to the quadratic equation $ax^2 + bx + c = 0$.

Now the system is: *ac* = 6, *ad* + *bc* = 23, and *bd* = 20, where *a* and *c* are divisors of 6, and *b* and *d* are divisors of 20.

That is to say that *a* and *c* are integers the product of which is 6, *b* and *d* are integers the product of which is 20, and *ad* + *bc* = 23.

And we can get the integers that can be *a*, *b*, *c*, and *d* via trial and error.

Suppose first, *a* = 1, *c* = 6, *b* = 4, and *d* = 5.
Then, *ad* + *bc* = 1·5 + 4·6 = 29, which is not 23, so we want to make another choice.

Suppose next, *a* = 1, *c* = 6, *b* = 5, and *d* = 4.
Then, *ad* + *bc* = 1·4 + 5·6 = 34, which is not 23, so we want to make another choice.

Suppose next, *a* = 2, *c* = 3, *b* = 5, and *d* = 4.
Then, *ad* + *bc* = 2·4 + 5·3, which is 23, so this time, we've got the right choice.

Now, we have: $P = 6x^2 + 23x + 20 = acx^2 + (ad + bc)x + bd = (ax + b)(cx + d)$, in which $a = 2$, $c = 3$, $b = 5$, and $d = 4$.

So we get: $P = 6x^2 + 23x + 20 = (ax + b)(cx + d) = (2x + 5)(3x + 4)$.

And each of $(2x + 5)$ and $(3x + 4)$ is prime.
If an integer or polynomial is prime, it has no divisor other than 1 and itself.
Therefore, $P = 6x^2 + 23x + 20$ is (fully) factorized to $(2x + 5)(3x + 4)$.

So such a factorization procedure looks quite complicated, doesn't it?
Don't get discouraged though. The entire process can take time and effort, but it is often the case the right choice of the integers will be made far much more quickly than we think. It's just a matter of practice.

Now, what if we cannot find the integers that can be a, b, c, and d anyway?

It can happen, of course. In that case, we want to consult the quadratic formula.
Using the formula, we can get the solution to:

$\frac{1}{6}(6x^2 + 23x + 20) = 0$, which is: $\frac{1}{6}P = 0$, which is: $x^2 + \frac{23}{6}x + \frac{20}{6} = 0$.

And the solution is: $x = -\frac{5}{2}$ or $-\frac{4}{3}$, which is the solution to $(x + \frac{5}{2})(x + \frac{4}{3}) = 0$.

Note that the solution above is to an equation $x^2 + \frac{23}{6}x + \frac{20}{6} = 0$, which is $\frac{1}{6}P = 0$.

Thus, we get:

$(x + \frac{5}{2})(x + \frac{4}{3}) = \frac{1}{6}P$, so $P = 6(x + \frac{5}{2})(x + \frac{4}{3}) = 2(x + \frac{5}{2})3(x + \frac{4}{3}) = (2x + 5)(3x + 4)$.

In short:

$6x^2 + 23x + 20 = 2{\cdot}3x^2 + (2{\cdot}4 + 3{\cdot}5)x + 4{\cdot}5 = 2{\cdot}3x^2 + 2{\cdot}4x + 3{\cdot}5x + 4{\cdot}5$

$= 2x(3x + 4) + 5(3x + 4) = (3x + 4)(2x + 5)$.

Suggestions or Solutions
To the **Problem** in the Example **1**

We have: $28x^2 + 71x + 18$.

Factorizing it, we can get it the way below:

$$28x^2 + 71x + 18 = 4 \cdot 7x^2 + (7 \cdot 9 + 2 \cdot 4)x + 2 \cdot 9 = 4 \cdot 7x^2 + 7 \cdot 9x + 2 \cdot 4x + 2 \cdot 9$$

$$= 7x(4x + 9) + 2(4x + 9) = (4x + 9)(7x + 2).$$

If not quite sure of the idea behind the processes above, follow the steps below:

Setting first, $P = 28x^2 + 71x + 18$, we can say that the polynomial P takes the form of a polynomial $ux^2 + vx + w$, which is similar to $acx^2 + (ad + bc)x + bd$, which is factorize to: $(ax + b)(cx + d)$.

So taking ac as u, $ad + bc$ as v, and bd as w, we get: $ux^2 + vx + w$, which P is in the form of. So P can be said to be similar to $acx^2 + (ad + bc)x + bd$, factorize to $(ax + b)(cx + d)$.

And we want a, b, c, and d to be integers.
That is to say that we want to factorize P within the scope of integers.

Assuming thus, a, b, c, and d are integers, setting: $ac = 28$, $bd = 18$, and $ad + bc = 71$, and then, finding a, b, c, and d, we can get the factorization of P.

So we want to solve a system where: $ac = 28$, $ad + bc = 71$, and $bd = 18$, where a and c are divisors of 28, and b and d are divisors of 18.

That is to say that a and c are integers the product of which is 28, b and d are integers the product of which is 18, and $ad + bc = 71$.

So a and c are divisors of 28, and b and d are divisors of 18.

And if any, we can get the integers that can be *a*, *b*, *c*, and *d* via trial and error.

Suppose first, $a = 1$, $c = 28$, $b = 2$, and $d = 9$.
Then, $ad + bc = 1 \cdot 9 + 2 \cdot 28 = 65$, which is not 71, so we want to make another choice.

Suppose next, $a = 2$, $c = 14$, $b = 3$, and $d = 6$.
Then, $ad + bc = 2 \cdot 6 + 3 \cdot 14 = 54$, which is not 71, so we want to make another choice.

Suppose next, $a = 7$, $c = 4$, $b = 2$, and $d = 9$.
Then, $ad + bc = 7 \cdot 9 + 2 \cdot 4$, which is 71, so this time, we've made the right choice.

Now, we have: $P = 28x^2 + 71x + 18 = acx^2 + (ad + bc)x + bd = (ax + b)(cx + d)$, where $a = 7$, $c = 4$, $b = 2$, and $d = 9$.

So we get: $P = 28x^2 + 71x + 18 = (ax + b)(cx + d) = (7x + 2)(4x + 9)$.

And each of $(7x + 2)$ and $(4x + 9)$ is prime.
If an integer or polynomial is prime, it has no divisor other than 1 and itself.

Therefore, $P = 28x^2 + 71x + 18$ is (fully) factorized to $(7x + 2)(4x + 9)$.

In short:

$28x^2 + 71x + 18 = 4 \cdot 7x^2 + (7 \cdot 9 + 2 \cdot 4)x + 2 \cdot 9 = 4 \cdot 7x^2 + 7 \cdot 9x + 2 \cdot 4x + 2 \cdot 9$

$= 7x(4x + 9) + 2(4x + 9) = (4x + 9)(7x + 2)$.

Examples C

Factorize the polynomials below.

0. $40t^2 + 59t + 21$

1. $30t^2 + 57t + 18$

2. $6r^2 - 13r - 28$

3. $\frac{3}{2}w^2 + 12w - 30$

4. $0.6t^2 + 14.92t - 2$

Suggestions or Solutions
To the **Problem** in the Example **0**

We have: $40t^2 + 59t + 21$.

Factorizing it, we can get it the way below:

$$40t^2 + 59t + 21 = 5 \cdot 8t^2 + (3 \cdot 8 + 7 \cdot 5)t + 3 \cdot 7 = 5 \cdot 8t^2 + 3 \cdot 8t + 7 \cdot 5t + 3 \cdot 7$$

$$= 8t(5t + 3) + 7(5t + 3) = (5t + 3)(8t + 7).$$

If not quite sure of the idea behind the processes above, follow the steps below:

Setting first, $P = 40t^2 + 59t + 21$, we can say that the polynomial P takes the form of a polynomial $ut^2 + vt + w$, which is similar to $act^2 + (ad + bc)t + bd$, which is factorize to $(at + b)(ct + d)$.

So taking ac as u, $ad + bc$ as v, and bd as w, we get: $ut^2 + vt + w$, which P is in the form of.

So P can be said to be similar to $act^2 + (ad + bc)t + bd$, factorize to $(at + b)(ct + d)$.

Assuming thus, a, b, c, and d are integers, setting $ac = 40$, $ad + bc = 59$, and $bd = 21$, and then, finding a, b, c, and d, we can get the factorization of P.

That is to say that a and c are integers the product of which is 40, b and d are integers the product of which is 21, and $ad + bc = 59$.

So a and c are divisors of 40, and b and d are divisors of 21.

And if any, we can get the integers that can be a, b, c, and d via trial and error.

Suppose first, $a = 1$, $c = 40$, $b = 3$, and $d = 7$.

Then, $ad + bc = 1 \cdot 7 + 3 \cdot 40 = 127$, which is not 59, so we want to make another choice.

Suppose next, $a = 4$, $c = 10$, $b = 3$, and $d = 7$.

Then, $ad + bc = 4 \cdot 7 + 3 \cdot 10 = 58$, which is not 59, so we want to make another choice.

Suppose next, $a = 5$, $c = 8$, $b = 3$, and $d = 7$.

Then, $ad + bc = 5 \cdot 7 + 3 \cdot 8$, which is 59, so we've got the right choice.

Now, we have: $P = 40t^2 + 59t + 21 = act^2 + (ad + bc)t + bd = (at + b)(ct + d)$, where $a = 5$, $c = 8$, $b = 3$, and $d = 7$.

So we get: $P = 40t^2 + 59t + 21 = (at + b)(ct + d) = (5t + 3)(8t + 7)$.

And each of $(5t + 3)$ and $(8t + 7)$ is prime.

Therefore, $P = 40t^2 + 59t + 21$ is fully factorized to $(5t + 3)(8t + 7)$.

In short:

$40t^2 + 59t + 21 = 5 \cdot 8t^2 + (3 \cdot 8 + 7 \cdot 5)t + 3 \cdot 7 = 5 \cdot 8t^2 + 3 \cdot 8t + 7 \cdot 5t + 3 \cdot 7$

$= 8t(5t + 3) + 7(5t + 3) = (5t + 3)(8t + 7)$.

Suggestions or Solutions

We have: $30t^2 + 57t + 18$.

Factorizing it, we can get it the way below:

To begin with, we get: $30t^2 + 57t + 18 = 3(10t^2 + 19t + 6)$.

And we get: $10t^2 + 19t + 6 = 2 \cdot 5t^2 + (3 \cdot 5 + 2 \cdot 2)t + 2 \cdot 3 = 2 \cdot 5t^2 + 3 \cdot 5t + 2 \cdot 2t + 2 \cdot 3$

$= 5t(2t + 3) + 2(2t + 3) = (2t + 3)(5t + 2)$.

So we get: $3(10t^2 + 19t + 6) = 3(2t + 3)(5t + 2)$.

Thus, we get: $30t^2 + 57t + 18 = 3(2t + 3)(5t + 2)$.

If not quite sure of the idea behind the processes above, follow the steps below:

Setting first, $P = 30t^2 + 57t + 18$, we can notice that 3 is a divisor common to every term in the polynomial P.

So we get: $P = 30t^2 + 57t + 18 = 3(10t^2 + 19t + 6)$.

So setting next, $Q = 10t^2 + 19t + 6$, we can set: $P = 3Q$, and Q is similar to a polynomial $act^2 + (ad + bc)t + bd$, which is factorize to $(at + b)(ct + d)$.

So putting Q in the form of $act^2 + (ad + bc)t + bd$, we can factorize Q to $(at + b)(ct + d)$.

Thus, finding a, b, c, and d, we can get the factorization of Q, and in turn, that of P's.

Since $Q = 10t^2 + 19t + 6$ takes the form of $act^2 + (ad + bc)t + bd$, we can set up a system of equations as follows: $ac = 10$, $ad + bc = 19$, and $bd = 6$.

Assuming thus, a, b, c, and d are integers, setting $ac = 10$, $ad + bc = 19$, and $bd = 6$, and then, finding a, b, c, and d, we can get the factorization of Q within the scope of integers.

That is to say that a and c are integers the product of which is 10, b and d are integers the product of which is 6, and $ad + bc = 19$.

So a and c are divisors of 40, and b and d are divisors of 21.

And if any, we can get the integers that can be a, b, c, and d via trial and error.

In fact, we get: $a = 2$, $c = 5$, $b = 3$, and $d = 2$, because we can get: $ac = 10$, $bd = 6$, and $ad + bc = 2 \cdot 2 + 3 \cdot 5$, which is 19.

Now, we have: $Q = 10t^2 + 19t + 6 = act^2 + (ad + bc)t + bd = (at + b)(ct + d)$, where $a = 2$, $c = 5$, $b = 3$, and $d = 2$.

Thus, we get: $Q = 10t^2 + 19t + 6 = (2t + 3)(5t + 2)$, so $P = 3Q = 3(2t + 3)(5t + 2)$.

And each of 3, $2t + 3$, and $5t + 2$ is prime.

Therefore, $P = 30t^2 + 57t + 18$ is (fully) factorized to $3(2t + 3)(5t + 2)$.

In short:

To begin with, we get: $30t^2 + 57t + 18 = 3(10t^2 + 19t + 6)$.

And we get: $10t^2 + 19t + 6 = 2 \cdot 5t^2 + (3 \cdot 5 + 2 \cdot 2)t + 2 \cdot 3 = 2 \cdot 5t^2 + 3 \cdot 5t + 2 \cdot 2t + 2 \cdot 3$

$= 5t(2t + 3) + 2(2t + 3) = (2t + 3)(5t + 2)$.

So we get: $3(10t^2 + 19t + 6) = 3(2t + 3)(5t + 2)$.

Thus, we get: $30t^2 + 57t + 18 = 3(2t + 3)(5t + 2)$.

Suggestions or Solutions
To the **Problem** in the Example **2**

We have: $6r^2 - 13r - 28$.

Factorizing it, we can get it the way below:

$6r^2 - 13r - 28 = 2 \cdot 3r^2 + (8 - 21)r + 4 \cdot 7 = 2 \cdot 3r^2 + (2 \cdot 4 - 3 \cdot 7)r + 4 \cdot 7$

$= 2 \cdot 3r^2 - 3 \cdot 7r + 2 \cdot 4r + 4 \cdot 7 = 3r(2r - 7) + 4(2r - 7) = (2r - 7)(3r + 4)$.

If not quite sure of the idea behind the processes above, follow the steps below:

Setting first, $P = 6r^2 - 13r - 28$, we can say that the polynomial P is similar to a polynomial $acr^2 + (ad + bc)r + bd$, which is factorize to $(ar + b)(cr + d)$.

So putting P in the form of $acr^2 + (ad + bc)r + bd$, we can factorize P to a polynomial in the form of $(ar + b)(cr + d)$.

Thus, finding a, b, c, and d, we can get the factorization of P.

Since $P = 6r^2 - 13r - 28$ takes the form of $acr^2 + (ad + bc)r + bd$, we can set up a system of equations as follows: $ac = 6$, $ad + bc = -13$, and $bd = -28$.

Assuming thus, a, b, c, and d are integers, setting $ac = 6$, $ad + bc = -13$, and $bd = -28$, and then, finding a, b, c, and d, we can get the factorization of P within the scope of integers.

That is to say that a and c are integers the product of which is 6, b and d are integers the product of which is -28, and $ad + bc = -13$.

So **a** and **c** are divisors of 6, and **b** and **d** are divisors of -28.

And if any, we can get the integers that can be **a**, **b**, **c**, and **d** via trial and error.

In fact, we get: $a = 2$, $c = 3$, $b = -7$, and $d = 4$, because we can get: $ac = 6$, $bd = -28$, and $ad + bc = 2 \cdot 4 + (-7 \cdot 3)$, which is -13.

Now, we have: $P = 6r^2 - 13r - 28 = acr^2 + (ad + bc)r + bd = (ar + b)(cr + d)$, in which $a = 2$, $c = 3$, $b = -7$, and $d = 4$.

Thus, we get: $P = 6r^2 - 13r - 28 = (2r - 7)(3r + 4)$.

Therefore, $P = 6r^2 - 13r - 28$ is factorized to $(2r - 7)(3r + 4)$.

In short:

$6r^2 - 13r - 28 = 2 \cdot 3r^2 + (8 - 21)r + 4 \cdot 7 = 2 \cdot 3r^2 + (2 \cdot 4 - 3 \cdot 7)r + 4 \cdot 7$

$= 2 \cdot 3r^2 - 3 \cdot 7r + 2 \cdot 4r + 4 \cdot 7 = 3r(2r - 7) + 4(2r - 7) = (2r - 7)(3r + 4)$.

Suggestions or Solutions
To the **Problem** in the Example **3**

We have: $\frac{3}{2}w^2 + 12w - 30$.

Factorizing it, we can get it the way below:

$$2P = 2 \cdot 3(\tfrac{1}{2}w^2 + 4w - 10) = 3(w^2 + 8w - 20) = 3(w-2)(w+10).$$

Thus, $P = \frac{3}{2}(w-2)(w+10)$.

If not quite sure of the idea behind the processes above, follow the steps below:

Let's set first, $P = \frac{3}{2}w^2 + 12w - 30$.

Unlike the previous ones, this one has is a fractional coefficient.
So what are we going to do about that?

Just get rid of the denominator.   How?

We can't just erase it with eraser, of course.
Multiplying by the denominator both sides of the equal sign, we can get rid of it.
Prior to the multiplication though, we want to take out 3 from the right hand side, first,
since 3 is a factor.

Then, we get: $2P = 2 \cdot 3(\tfrac{1}{2}w^2 + 4w - 10) = 3(w^2 + 8w - 20) = 3(w-2)(w+10)$.

So we get: $P = \frac{3}{2}(w-2)(w+10)$.   And each of $(w-2)$ and $(w+10)$ is prime.

Thus, P is factorized to $\frac{3}{2}(w-2)(w+10)$.

Suggestions or Solutions
To the **Problem** in the Example **4**

We have: $0.6t^2 + 14.92t - 2$.

Factorizing it, we can get it the way below:

To begin with, $0.6t^2 + 14.92t - 2 = 0.01(60t^2 + 1492t - 200) = 0.04(15t^2 + 373t - 50)$.

And we get: $15t^2 + 373t - 50 = 1 \cdot 15t^2 + (375 - 2)t + 2 \cdot 25 = 1 \cdot 15t^2 + (15 \cdot 25 - 1 \cdot 2)t + 2 \cdot 25$

$= 1 \cdot 15t^2 + 15 \cdot 25t - 1 \cdot 2t + 2 \cdot 25 = 15t(t + 25) - 2(t + 25) = (t + 25)(15t - 2)$.

So we get: $0.04(15t^2 + 373t - 50) = 0.04(t + 25)(15t - 2)$, which can be now, taken as the factorization of $0.6t^2 + 14.92t - 2$.

And if necessary, we can factorize the coefficient 0.04, too. We have: $0.04 = 2^2 10^{-2}$.

So we can put it this way, too: $0.6t^2 + 14.92t - 2 = 2^2 10^{-2}(t + 25)(15t - 2)$.

If not quite sure of the idea behind the processes above, follow the steps below:

Let's set first, $P = 0.6t^2 + 14.92t - 2$.
This one has coefficients fractional, and this time, they are decimal fractional.
So what are we going to do about those?

Get rid of them, of course. Multiplying by 100 both sides of the equal sign, we get:

$100P = 100(0.6t^2 + 14.92t - 2) = 60t^2 + 1492t - 200$, which can be reduced a bit further to $4(15t^2 + 373t - 50)$.   So we now want to factorize $15t^2 + 373t - 50$.

18

So next, setting: $Q = 15t^2 + 373t - 50$, we get: $100P = 4Q$, and Q is similar to a polynomial $act^2 + (ad + bc)t + bd$, which is factorize to $(at + b)(ct + d)$.

Since $Q = 15t^2 + 373t - 50$ takes the form of $act^2 + (ad + bc)t + bd$, we can set up a system of equations as follows: $ac = 15$, $ad + bc = 373$, and $bd = -50$.

Assuming thus, a, b, c, and d are integers, setting $ac = 15$, $ad + bc = 373$, and $bd = -50$, and then, finding a, b, c, and d, we can get the factorization of Q within the scope of integers.

That is to say that a and c are integers the product of which is 15, b and d are integers the product of which is -50, and $ad + bc = 373$.

So a and c are divisors of 15, and b and d are divisors of -50.
And if any, we can get the integers that can be a, b, c, and d via trial and error.

In fact, we get: $a = 1$, $c = 15$, $b = 25$, and $d = -2$, because we can get: $ac = 15$, $bd = -50$, and $ad + bc = 1 \cdot (-2) + 25 \cdot 15 = -2 + 25(10 + 5) = -2 + 250 + 125 = 373$.

Now, we have: $Q = 15t^2 + 373t - 50 = act^2 + (ad + bc)t + bd = (at + b)(ct + d)$, in which $a = 1$, $c = 15$, $b = 25$, and $d = -2$.

Thus, we get $Q = 15t^2 + 373t - 50 = (t + 25)(15t - 2)$, so $100P = 4Q = 4(t + 25)(15t - 2)$.

And each of 4, $t + 25$, and $15t - 2$ is prime.   And we have: $0.04 = 2^2 10^{-2}$.

Therefore, $P = 0.6t^2 + 14.92t - 2$ is factorized to $2^2 10^{-2}(t + 25)(15t - 2)$.

And of course, we can just put it this way, too:

$0.6t^2 + 14.92t - 2 = 0.04(t + 25)(15t - 2)$.

Examples D

Factorize the polynomials below.

0. $x^3 + 10x^2 + 31x + 30$

1. $x^3 - 10x^2 + 31x - 30$

2. $x^3 + 4x^2 - x - 4$

Suggestions or Solutions
To the **Problem** in the Example **0**

We have: $x^3 + 10x^2 + 31x + 30$.

Factorizing it, we can get it the way below:

Setting first, $P = x^3 + 10x^2 + 31x + 30$, and evaluating P for $x = -2$, we get:

$(-2)^3 + 10(-2)^2 + 31(-2) + 30 = -8 + 40 - 62 + 30 = 0$.

So $x + 2$ is a divisor, that is, a factor of P.

Thus next, applying to P the synthetic division by $(x + 2)$, we get:

-2	1	10	31	30
		-2	-16	-30
	1	8	15	0

So we get: $P = x^3 + 10x^2 + 31x + 30 = (x + 2)(x^2 + 8x + 15) = (x + 2)(x + 3)(x + 5)$.

If not quite sure of the idea behind the processes above, follow the steps below:

Let's set first, $P = x^3 + 10x^2 + 31x + 30$.

Unlike the polynomials in the previous examples, the polynomial P is of degree 3, and doesn't seem to have any divisor that we can quickly see.

If P can get factorized though, it will probably be a product of polynomials of degree 2 or 1. And we have covered factorizations of polynomials of degree 2.

So let's go over factorizations of polynomials of degree 2.

To begin with, a polynomial $x^2 + 2xy + y^2$ gets factorized to $(x + y)^2$, which is called a complete square.

Next, $x^2 + ux + v$ can be factorized to $(x + m)(x + n)$ where $u = m + n$ and $v = mn$. Next, $sx^2 + tx + w$ can be factorized to $(ax + b)(cx + d)$ where $s = ac$, $t = ad + bc$, and $w = bd$.

So we can see a pattern that a polynomial of degree 2 can get factorized to a product of two factors, each of which is of degree 1.

What then, about the factorizations of a polynomial of degree 3?

We can expect that a polynomial of degree 3 can get factorized to the form of a polynomial as $(x + a)(x + b)(x + c)$.

More specifically, we can expect the polynomial to get factorized first, to a product of a polynomial of degree 1 as $x + d$ and a polynomial of degree 2 as $x^2 + ux + v$, which in tern, can be factorized to $(x + m)(x + n)$, so eventually, the polynomial of degree 3 can be factorized to $(x + a)(x + b)(x + c)$.

In fact, we have a factorization identity where $x^3 + (a + b + c)x^2 + (ab + bc + ca)x + abc$

$= (x + a)(x + b)(x + c)$.

So if factorizing a polynomial in the form of $x^3 + ux^2 + vx + w$, assuming that it can be factorized to a form of $(x + a)(x + b)(x + c)$, we can try finding the values of a, b, and c.

How come?

We know that the polynomial P is similar to $x^3 + ux^2 + vx + w$, which is again, similar to $x^3 + (a + b + c)x^2 + (ab + bc + ca)x + abc$, which gets factorized to $(x + a)(x + b)(x + c)$.

So we can expect that P gets factorized to $(x + a)(x + b)(x + c)$, where a, b, and c are integers if P gets factorized within the scope of integers.

Then, $(x + a)$, $(x + b)$, and $(x + c)$ are factors of P.

If setting thus, $P = x^3 + 10x^2 + 31x + 30 = (x + a)(x + b)(x + c)$, we can try finding a, b, and c, assuming they are integers, i.e., P gets factorized within the scope of integers.

How can we find them though?

Expanding $(x + a)(x + b)(x + c)$, we get: $x^3 + (a + b + c)x^2 + (ab + bc + ca)x + abc$.

And we have set: $x^3 + 10x^2 + 31x + 30 = (x + a)(x + b)(x + c)$.

So we can set: $P = x^3 + 10x^2 + 31x + 30 = x^3 + (a + b + c)x^2 + (ab + bc + ca)x + abc$.

Comparing thus, both sides of the equality above term by term, we get: $30 = abc$. So we can notice that a, b, and c can be divisors of 30, and so can the negatives of a, b, and c.

Now, getting back to $x^3 + 10x^2 + 31x + 30 = (x + a)(x + b)(x + c)$, we can see that:

$(-a)^3 + 10(-a)^2 + 31(-a) + 30 = (-a + a)(-a + b)(-a + c) = 0$ if $x = -a$.

$(-b)^3 + 10(-b)^2 + 31(-b) + 30 = (-b + a)(-b + b)(-b + c) = 0$ if $x = -b$.

$(-c)^3 + 10(-c)^2 + 31(-c) + 30 = (-c + a)(-c + b)(-c + c) = 0$ if $x = -c$.

And we assume that a, b, and c are divisors of 30, and so are the negatives of a, b, and c.

So putting a divisor of 30 into x in P, and getting: $P = 0$, we can say that the ***negative*** of the divisor is one of a, b, and c.

For instance, we can take it as a. Then, $x + a$ is a factor of P, since it divides P, and it is in fact, one of the three factors of P.

And the three are: $(x + a)$, $(x + b)$, and $(x + c)$, since each divides P.

How then, can we get the other two factors?

We know: $(x + a)$, $(x + b)$, and $(x + c)$ are factors of P.

And factors are divisors. So $x + a$ can divide P. What then, is the quotient?

We have: $P = x^3 + 10x^2 + 31x + 30 = (x + a)(x + b)(x + c)$.

So dividing P by $x + a$, we get: $(x + b)(x + c)$, which is thus, the quotient.

The quotient will not be however, in the form of $(x + b)(x + c)$, of course.
It will be in the form of $x^2 + mx + n$, which will be though, factorized to $(x + b)(x + c)$.

So let's now get the factorization of P using the method above.

First, the divisors of 30 are: ± 1, ± 2, ± 3, ± 6, ± 10, ± 15, and ± 30.

Next, putting a divisor of 30 into x in $x^3 + 10x^2 + 31x + 30$, and getting 0, we can say that the *negative* of the divisor is the value of one of a, b, and c.

So if P can be factorized within the scope of integers, we will get to find an integer that makes P be **0**, and then, the integer is one of the divisors of 30 in this case, and the *negative* of the integer is one of a, b, and c.

Now, evaluating $P = x^3 + 10x^2 + 31x + 30$ for $x = $ **-2**, we get:
$(-2)^3 + 10(-2)^2 + 31(-2) + 30 = -8 + 40 - 62 + 30 = 0$.

We know that we have set: $P = x^3 + 10x^2 + 31x + 30 = (x + a)(x + b)(x + c)$.
So at least one of a, b, and c is 2.

And thus, we can now see that $x + 2$ is a factor of $P = x^3 + 10x^2 + 31x + 30$.
So next, applying to P the synthetic division by $(x + 2)$, we get:

-2	1	10	31	30
		-2	-16	-30
	1	8	15	0

Thus, dividing $x^3 + 10x^2 + 31x + 30$ by $x + 2$, we get: $x^2 + 8x + 15$ as the quotient.

So we get: $P = x^3 + 10x^2 + 31x + 30 = (x + 2)(x^2 + 8x + 15)$.

And thus, $x + 2$ is a factor of P, and is one of the three factors of P.

What then, is the quotient $x^2 + 8x + 15$?

The quotient $x^2 + 8x + 15$ is the product of the other two factors of P.

And in fact, $x^2 + 8x + 15$ is factorized to $(x + 3)(x + 5)$.

So we get: $P = x^3 + 10x^2 + 31x + 30 = (x + 2)(x^2 + 8x + 15) = (x + 2)(x + 3)(x + 5)$.

And each of $x + 2$, $x + 3$, and $x + 5$ is prime.

Therefore, $P = x^3 + 10x^2 + 31x + 30$ gets (fully) factorized to $(x + 2)(x + 3)(x + 5)$.

In short:

Setting first, $P = x^3 + 10x^2 + 31x + 30$, and evaluating P for $x = -2$, we get:

$(-2)^3 + 10(-2)^2 + 31(-2) + 30 = -8 + 40 - 62 + 30 = 0$.

So $x + 2$ is a divisor, that is, a factor of P.

Thus next, applying to P the synthetic division by $(x + 2)$, we get:

$$
\begin{array}{r|rrrr}
-2 & 1 & 10 & 31 & 30 \\
   &   & -2 & -16 & -30 \\
\hline
   & 1 & 8 & 15 & 0
\end{array}
$$

So we get: $P = x^3 + 10x^2 + 31x + 30 = (x + 2)(x^2 + 8x + 15) = (x + 2)(x + 3)(x + 5)$.

Suggestions or Solutions
To the **Problem** in the Example **1**

We have: $x^3 - 10x^2 + 31x - 30$.

Factorizing it, we can get it the way below:

Setting first, $P = x^3 - 10x^2 + 31x - 30$, and evaluating P for $x = 2$, we get:

$2^3 - 10 \cdot 2^2 + 31 \cdot 2 - 30 = 8 - 40 + 62 - 30 = 0$.

So $x - 2$ is a divisor, that is, a factor of $P = x^3 - 10x^2 + 31x - 30$.

Next, applying to P the synthetic division by $(x - 2)$, we get:

2	1	-10	31	-30
		2	-16	30
	1	-8	15	0

Thus, we get: $x^3 - 10x^2 + 31x - 30 = (x - 2)(x^2 - 8x + 15) = (x - 2)(x - 3)(x - 5)$.

If not quite sure of the idea behind the processes above, follow the steps below:

Setting first, $P = x^3 - 10x^2 + 31x - 30$, we can say that the polynomial P is similar to $x^3 + ux^2 + vx + w$, which is again, similar to $x^3 + (a + b + c)x^2 + (ab + bc + ca)x + abc$, which gets factorized to $(x + a)(x + b)(x + c)$.

So we can expect that P gets factorized to $(x + a)(x + b)(x + c)$, where a, b, and c are integers.

Setting thus, $P = x^3 - 10x^2 + 31x - 30 = (x + a)(x + b)(x + c)$, we can try finding a, b, and c, assuming they are integers, i.e., P gets factorized within the scope of integers.

How can we find them though?

Expanding $(x + a)(x + b)(x + c)$, we get: $x^3 + (a + b + c)x^2 + (ab + bc + ca)x + abc$.

And we have set: $P = x^3 - 10x^2 + 31x - 30 = (x + a)(x + b)(x + c)$.

So we can set: $P = x^3 - 10x^2 + 31x - 30 = x^3 + (a + b + c)x^2 + (ab + bc + ca)x + abc$.

Comparing therefore, both sides of the equal sign term by term, we get: $-30 = abc$. So we can say that a, b, and c can be divisors of -30, and so can the negatives of a, b, and c.

Now, getting back to $x^3 - 10x^2 + 31x - 30 = (x + a)(x + b)(x + c)$, we can see that:

$(-a)^3 - 10(-a)^2 + 31(-a) - 30 = (-a + a)(-a + b)(-a + c) = 0$ if $x = -a$.

$(-b)^3 - 10(-b)^2 + 31(-b) - 30 = (-b + a)(-b + b)(-b + c) = 0$ if $x = -b$.

$(-c)^3 - 10(-c)^2 + 31(-c) - 30 = (-c + a)(-c + b)(-c + c) = 0$ if $x = -c$.

And we assume that a, b, and c are divisors of -30, and so are the negatives of a, b, and c.

So putting a divisor of -30 into x in P, and getting: $P = 0$, we can say that the negative of the divisor is one of a, b, and c. For instance, we can take it as a.

How then, can we get the other two integers b and c?

We know: $(x + a)$, $(x + b)$, and $(x + c)$ are factors of P.

And factors are divisors.   So $x + a$ can divide P.   What then, is the quotient?

We have: $P = x^3 - 10x^2 + 31x - 30 = (x + a)(x + b)(x + c)$.

So dividing P by $x + a$, we get: $(x + b)(x + c)$, which is thus, the quotient.

The quotient will be however, in the form of $x^2 + mx + n$, which will be though, factorized to $(x + b)(x + c)$.

So let's now get the factorization of P using the method above.

First, the divisors of -30 are: ± 1, ± 2, ± 3, ± 6, ± 10, ± 15, and ± 30.

Next, putting one of the divisors into x in $x^3 - 10x^2 + 31x - 30$, and getting 0, we can say that the one is the value of one of a, b, and c.

So if P can be factorized within the scope of integers, we will get to find an integer that can give us $P = 0$, and the integer is one of the divisors of -30 in this case.

Now, evaluating $P = x^3 - 10x^2 + 31x - 30$ for $x = 2$, we get:

$2^3 - 10 \cdot 2^2 + 31 \cdot 2 - 30 = 8 - 40 + 62 - 30 = 0$.

And we know that we have set: $P = x^3 - 10x^2 + 31x - 30 = (x + a)(x + b)(x + c)$.

So at least one of a, b, and c is -2.

And thus, we can now see that $x - 2$ is a factor of $P = x^3 - 10x^2 + 31x - 30$.

So next, applying to P the synthetic division by $(x - 2)$, we get:

```
2 | 1     -10    31    -30
  |         2   -16     30
  ‾‾‾‾‾‾‾‾‾‾‾‾‾‾‾‾‾‾‾‾‾‾‾‾
    1      -8    15      0
```

Thus, dividing $x^3 - 10x^2 + 31x - 30$ by $x - 2$, we get: $x^2 - 8x + 15$ as the quotient.

So we get: $x^3 - 10x^2 + 31x - 30 = (x - 2)(x^2 - 8x + 15)$.

Next, we want to factorize the quotient $x^2 - 8x + 15$.

In fact, $x^2 - 8x + 15$ is factorized to $(x - 3)(x - 5)$.

So we get: $x^3 - 10x^2 + 31x - 30 = (x - 2)(x^2 - 8x + 15) = (x - 2)(x - 3)(x - 5)$.

And each of $x - 2$, $x - 3$, and $x - 5$ is prime.

Therefore, $P = x^3 - 10x^2 + 31x - 30$ gets (fully) factorized to $(x - 2)(x - 3)(x - 5)$.

Notice that replacing -2, -3, and -5 above with 2, 3, and 5 respectively, and using the identity where $x^3 + (a + b + c)x^2 + (ab + bc + ca)x + abc = (x + a)(x + b)(x + c)$, we get:

$(x + 2)(x + 3)(x + 5) = x^3 + (2 + 3 + 5)x^2 + (6 + 15 + 10)x + 30 = x^3 + 10x^2 + 31x + 30$.

So we can see at once, $x^3 + 10x^2 + 31x + 30$ gets fully factorized to $(x + 2)(x + 3)(x + 5)$.

In short:

Setting first, $P = x^3 - 10x^2 + 31x - 30$, and evaluating P for $x = 2$, we get:

$2^3 - 10 \cdot 2^2 + 31 \cdot 2 - 30 = 8 - 40 + 62 - 30 = 0$.

So $x - 2$ is a divisor, that is, a factor of $P = x^3 - 10x^2 + 31x - 30$.

Next, applying to P the synthetic division by $(x - 2)$, we get:

2	1	-10	31	-30
		2	-16	30
	1	-8	15	0

Thus, we get: $x^3 - 10x^2 + 31x - 30 = (x - 2)(x^2 - 8x + 15) = (x - 2)(x - 3)(x - 5)$.

Suggestions or Solutions

To the **Problem** in the Example **2**

We have: $x^3 + 4x^2 - x - 4$.

Factorizing it, we can get it the way below:

Setting first, $P = x^3 + 4x^2 - x - 4$, and evaluating P for $x = 1$, we get: $1 + 4 - 1 - 4 = 0$, so $x - 1$ is a factor, and thus, applying to P the synthetic division by $x - 1$, we get:

1	1	4	-1	-4
		1	5	4
	1	5	4	0

Thus, we get: $x^3 + 4x^2 - x - 4 = (x - 1)(x^2 + 5x + 4) = (x - 1)(x + 1)(x + 4)$.

If not quite sure of the idea behind the processes above, follow the steps below:

Setting first, $P = x^3 + 4x^2 - x - 4$, we can say that the polynomial P is similar to the polynomial $x^3 + (a + b + c)x^2 + (ab + bc + ca)x + abc$, factorized to $(x + a)(x + b)(x + c)$.

So we can expect that P gets factorized to $(x + a)(x + b)(x + c)$, where a, b, and c are integers.

Setting thus, $P = x^3 + 4x^2 - x - 4 = (x + a)(x + b)(x + c)$, we can try finding a, b, and c, assuming they are integers, that is, P gets factorized within the scope of integers.

Next, we have: $(x + a)(x + b)(x + c) = x^3 + (a + b + c)x^2 + (ab + bc + ca)x + abc$.

And we have set: $P = x^3 + 4x^2 - x - 4 = (x + a)(x + b)(x + c)$.

So we can set: $P = x^3 + 4x^2 - x - 4 = x^3 + (a + b + c)x^2 + (ab + bc + ca)x + abc$.

Comparing therefore, both sides of the equal sign term by term, we get: $-4 = abc$. So we can say that a, b, and c can be divisors of -4, and so can the negatives of a, b, and c.

Now, getting back to $x^3 + 4x^2 - x - 4 = (x + a)(x + b)(x + c)$, we can see that:

$$(-a)^3 + 4(-a)^2 - (-a) - 4 = (-a + a)(-a + b)(-a + c) = 0 \text{ if } x = -a.$$

And we assume that a is a divisor of -4, and so is the negative of a.

So putting a divisor of -4 into x in P, and getting: $P = 0$, we can say that the negative of the divisor can be a.   How then, can we get the other two, which are b and c?

We know: $(x + a)$, $(x + b)$, and $(x + c)$ are factors of P.

And factors are divisors.   So $x + a$ can divide P.

And dividing P by $x + a$, we get: $(x + b)(x + c)$, which is thus, the quotient.

The quotient will be however, in the form of of $x^2 + mx + n$, which will be though, factorized to $(x + b)(x + c)$.

So let's now get the factorization of P using the method above.

First, the divisors of -4 are: ± 1, ± 2, and ± 4.

Next, evaluating $P = x^3 + 4x^2 - x - 4$ for $x = 1$, we get: $1 + 4 - 1 - 4 = 0$.

So $x - 1$ is a factor of P, and applying to P the synthetic division by $x - 1$, we get:

$$
\begin{array}{c|cccc}
1 & 1 & 4 & -1 & -4 \\
  &   & 1 & 5 & 4 \\
\hline
  & 1 & 5 & 4 & 0 \\
\end{array}
$$

So we get: $x^3 + 4x^2 - x - 4 = (x - 1)(x^2 + 5x + 4) = (x - 1)(x + 1)(x + 4)$.

And each of $x - 1$, $x + 1$, and $x + 4$ is prime.

Therefore, $P = x^3 + 4x^2 - x - 4$ gets fully factorized to $(x - 1)(x + 1)(x + 4)$.

Examples E

Factorize the polynomials below.

0. $x^3 + 2x^2 - 7x + 4$

1. $x^3 - 3x^2 + 3x - 1$.

2. $6x^3 - 5x^2 - 12x - 4$

Suggestions or Solutions
To the **Problem** in the Example **0**

We have: $x^3 + 2x^2 - 7x + 4$.

Factorizing it, we can get it the way below:

Setting first, $P = x^3 + 2x^2 - 7x + 4$, and evaluating P for $x = 1$, we get: $1 + 2 - 7 + 4 = 0$, so $x - 1$ is a factor, and thus, applying to P the synthetic division by $x - 1$, we get:

1 | 1   2   -7   4
  | 1   3   -4
  ----
    1   3   -4   0
```

Thus, we get:

$$x^3 + 2x^2 - 7x + 4 = (x - 1)(x^2 + 3x - 4) = (x - 1)(x - 1)(x + 4) = (x - 1)^2(x + 4).$$

*If not quite sure of the idea behind the processes above, follow the steps below:*

Setting first, $P = x^3 + 2x^2 - 7x + 4$, we can say that $P$ is similar to the polynomial $x^3 + (a + b + c)x^2 + (ab + bc + ca)x + abc$, factorized to $(x + a)(x + b)(x + c)$.

So we can expect that $P$ gets factorized to $(x + a)(x + b)(x + c)$, where $a$, $b$, and $c$ are integers.

Setting thus, $P = x^3 + 2x^2 - 7x + 4 = (x + a)(x + b)(x + c)$, we can try finding $a$, $b$, and $c$, assuming that they are integers, that is, $P$ gets factorized within the scope of integers.

Now, expanding $(x + a)(x + b)(x + c)$, we get: $x^3 + (a + b + c)x^2 + (ab + bc + ca)x + abc$.

So we can set: $P = x^3 + 2x^2 - 7x + 4 = x^3 + (a + b + c)x^2 + (ab + bc + ca)x + abc$.
```

Comparing therefore, the terms in both sides of the equal sign, we get: **4 = abc**. So we can assume that a, b, and c are divisors of 4, and so are the negatives of a, b, and c.

Now, getting back to $x^3 + 2x^2 - 7x + 4 = (x + a)(x + b)(x + c)$, we can see that:

$(-a)^3 + 2(-a)^2 - 7(-a) + 4 = (-a + a)(-a + b)(-a + c) = 0$ if $x = -a$.

And we assume that a is a divisor of 4, and so is $-a$.

So putting a divisor of 4 into x in P, and getting: $P = 0$, we can say that the negative of the divisor can be a. How then, can we get the other two, which are b and c?

We know: $(x + a)$, $(x + b)$, and $(x + c)$ are factors of P. So $x + a$ can divide P.

And dividing P by $x + a$, we get: $(x + b)(x + c)$, which is thus, the quotient.

The quotient will be however, in the form of of $x^2 + mx + n$, which will be though, factorized to $(x + b)(x + c)$.

So let's now get the factorization of P using the method above.

First, the divisors of 4 are: ± 1, ± 2, and ± 4.

Next, evaluating $P = x^3 + 2x^2 - 7x + 4$ for $x = 1$, we get: $1 + 2 - 7 + 4 = 0$.

So $x - 1$ is a factor of P, and applying to P the synthetic division by $x - 1$, we get:

$$
\begin{array}{c|cccc}
1 & 1 & 2 & -7 & 4 \\
  &   & 1 & 3 & -4 \\
\hline
  & 1 & 3 & -4 & 0 \\
\end{array}
$$

So we get:

$x^3 + 2x^2 - 7x + 4 = (x - 1)(x^2 + 3x - 4) = (x - 1)(x - 1)(x + 4) = (x - 1)^2(x + 4)$.

Suggestions or Solutions
To the Problem in the Example 1

We have: $x^3 - 3x^2 + 3x - 1$.

Factorizing it, we can get it the way below:

Setting first, $P = x^3 - 3x^2 + 3x - 1$, and evaluating P for $x = 1$, we get: $1 - 3 + 3 - 1 = 0$, so $x - 1$ is a factor, and thus, applying to P the synthetic division by $x - 1$, we get:

$$
\begin{array}{c|cccc}
1 & 1 & -3 & 3 & -1 \\
 & & 1 & -2 & 1 \\
\hline
 & 1 & -2 & 1 & 0
\end{array}
$$

Thus, we get: $x^3 - 3x^2 + 3x - 1 = (x - 1)(x^2 - 2x + 1) = (x - 1)(x - 1)^2 = (x - 1)^3$.

In fact, we have factorization identities as follows:

$x^3 + 3x^2y + 3xy^2 + y^3 = (x + y)^3$.

$x^3 - 3x^2y + 3xy^2 - y^3 = (x - y)^3$.

And putting both together, we can set: $x^3 \pm 3x^2y + 3xy^2 \pm y^3 = (x \pm y)^3$.

If not quite sure of the idea behind the processes above, follow the steps below:

Setting first, $P = x^3 - 3x^2 + 3x - 1$, we can say that the polynomial P is similar to $x^3 + (a + b + c)x^2 + (ab + bc + ca)x + abc$, which gets factorized to $(x + a)(x + b)(x + c)$.

So we can expect P to get factorized to $(x + a)(x + b)(x + c)$.

Setting thus, $P = x^3 - 3x^2 + 3x - 1 = (x + a)(x + b)(x + c)$, we can try finding a, b, and c, assuming they are integers, that is, P gets factorized within the scope of integers.

Now, expanding $(x + a)(x + b)(x + c)$, we get: $x^3 + (a + b + c)x^2 + (ab + bc + ca)x + abc$.

So we can set: $P = x^3 - 3x^2 + 3x - 1 = x^3 + (a + b + c)x^2 + (ab + bc + ca)x + abc$.

Comparing therefore, both sides of the equal sign term by term, we get: **-1 = abc**. So we can assume that *a*, *b*, and *c* are divisors of -1, and so are the negatives of *a*, *b*, and *c*.

Now, getting back to $x^3 - 3x^2 + 3x - 1 = (x + a)(x + b)(x + c)$, we can see that:

$(-a)^3 + 2(-a)^2 - 7(-a) + 4 = (-a + a)(-a + b)(-a + c) = 0$ if $x = -a$.

And we assume that *a* is a divisor of -1, and so is *-a*.

So putting a divisor of -1 into *x* in *P*, and getting: **P = 0**, we can say that the negative of the divisor can be *a*.

And we know: $(x + a)$, $(x + b)$, and $(x + c)$ are factors of *P*. So $x + a$ can divide *P*.

And dividing *P* by $x + a$, we get: $(x + b)(x + c)$, which is thus, the quotient.

The quotient will be however, in the form of of $x^2 + mx + n$, which will be though, factorized to $(x + b)(x + c)$.
So let's now get the factorization of *P* using the method above.

First, the divisors of -1 are: ± 1.

Next, evaluating $P = x^3 - 3x^2 + 3x - 1$ for **x = 1**, we get: $1 - 3 + 3 - 1 = 0$.

So **x – 1** is a factor of *P*, and applying to *P* the synthetic division by $x - 1$, we get:

```
1  |  1    -3     3    -1
   |        1    -2     1
   |_____
      1    -2     1     0
```

So we get:

$x^3 - 3x^2 + 3x - 1 = (x - 1)(x^2 - 2x + 1) = (x - 1)(x - 1)(x - 1) = (x - 1)^3$.

Suggestions or Solutions
To the **Problem** in the Example **2**

We have: $6x^3 - 5x^2 - 12x - 4$.

Factorizing it, we can get the way below:

Evaluating $6x^3 - 5x^2 - 12x - 4$ for $x = -\frac{2}{-1} = 2$, we get: $48 - 20 - 24 - 4 = 0$, so we get:

2	6	-5	-12	-4
		12	14	4
	6	7	2	0

Thus, we get: $6x^3 - 5x^2 - 12x - 4 = (x - 2)(6x^2 + 7x + 2) = (x - 2)(2x + 1)(3x + 2)$.

If not quite sure of the idea behind the processes above, follow the steps below:

Let's set first, $P = 6x^3 - 5x^2 - 12x - 4$.

The polynomial P is a bit different from the previous one.

Unlike the preceding examples, the coefficient in the highest term is not 1 but 6. So this time, it's not the case where we just go ahead and use the identity below:

$x^3 + (a + b + c)x^2 + (ab + bc + ca)x + abc = (x + a)(x + b)(x + c)$.

So let's now look for some other way we can try.

We know a polynomial $acx^2 + (ad + bc)x + bd$ gets factorize to $(ax + b)(cx + d)$.

So we can reasonably expect that a polynomial $sx^3 + tx^2 + ux + v$ can get factorized to $(ax + b)(cx + d)(ex + g)$.

In fact, expanding $(ax + b)(cx + d)(ex + g)$, we get:

$acex^3 + (acg + \text{etc.})x^2 + (adg + \text{etc.})x + bdg$, which is quite similar to $sx^3 + tx^2 + ux + v$.

So assuming $sx^3 + tx^2 + ux + v$ gets factorized to $(ax + b)(cx + d)(ex + g)$, we can set:

$sx^3 + tx^2 + ux + v = acex^3 + (acg + \text{etc.})x^2 + (adg + \text{etc.})x + bdg$.

What then, do we get?

We can get: $s = ace$, and $v = bdg$.   So?

Assuming a, c, and e are divisors of s, and b, d, and g are divisors of v, and finding a, b, c, d, e, and g by the similar method applied to $x^3 + ux^2 + vx + w$, we could get the factorization of $sx^3 + tx^2 + ux + v$.

How then, can we find a, b, c, d, e, and g?

Setting $A = sx^3 + tx^2 + ux + v = (ax + b)(cx + d)(ex + g)$, we can see that:

$A = 0$ when $x = -\frac{b}{a}$, $-\frac{d}{c}$, or $-\frac{g}{e}$.

Thus, if $m = \frac{b}{a}$, $\frac{d}{c}$, or $\frac{g}{e}$, we can use $x + m$ as a factor of A.   How come?

We can put $(ax + b)(cx + d)(ex + g)$ the way below:

$(ax + b)(cx + d)(ex + g) = a(x+\frac{b}{a})c(x+\frac{d}{c})e(x+\frac{g}{e}) = ace(x+\frac{b}{a})(x+\frac{d}{c})(x+\frac{g}{e})$.

So we can now set: $A = sx^3 + tx^2 + ux + v = ace(x + \frac{b}{a})(x + \frac{d}{c})(x + \frac{g}{e})$.

And thus, we can say that $(x + \frac{b}{a})$, $(x + \frac{d}{c})$, and $(x + \frac{g}{e})$ are factors of A.

Suppose now that we have found $\frac{b}{a}$, for instance.

Then, applying to A the synthetic division by $(x + \frac{b}{a})$, we get as the quotient a polynomial of degree 2, which is equivalent to $ace(x + \frac{d}{c})(x + \frac{g}{e})$, of course.

What then about the other two factors $(x + \frac{d}{c})$ and $(x + \frac{g}{e})$?

We have: $ace(x + \frac{d}{c})(x + \frac{g}{e}) = ac(x + \frac{d}{c})e(x + \frac{g}{e}) = a(cx + d)(ex + g)$.

And we know how to get $a(cx + d)(ex + g)$ as a factorization.

So we will eventually get: $(x + \frac{b}{a})a(cx + d)(ex + g) = (ax + b)(cx + d)(ex + g)$.

And thus, factorizing a polynomial as $sx^3 + tx^2 + ux + v$, we can begin with assuming:
$A = sx^3 + tx^2 + ux + v = (ax + b)(cx + d)(ex + g)$.   What then, is the next?

We want to find the value of x that makes $A = 0$. How then, can we find the value of x?

We know that setting $A = sx^3 + tx^2 + ux + v = (ax + b)(cx + d)(ex + g)$, we get:

$A = 0$ if $x = -\frac{b}{a}, -\frac{d}{c},$ or $-\frac{g}{e}$.   And we assume that $s = ace$, and $v = bdg$.
So first, we can assume that $a, c,$ and e are divisors of s, and $b, d,$ and g are divisors of v.

And by trial and error, coming up with, for instance, $-\frac{b}{a}$ that makes $A = 0$, we can get a factor of A.   And the factor is: $x + \frac{b}{a}$.

And next, we can get the other factors if factorizing the quotient we get after applying to A the synthetic division by the factor.

And of course, if the quotient is prime, we can use the quotient itself as the other factor.

So let's now, factorize $P = 6x^3 - 5x^2 - 12x - 4$ using the idea above.

Then, first, we want to find a value of x that makes $P = 0$. And the value is a fraction where the numerator is a divisor of -4, and the denominator is a divisor of 6.

(The divisors of 6 are ±1, ±2, ±3, and ±6, and the divisors of -4 are ±1, ±2, and ±4.)

So for instance, using as a numerator, 2 (a divisor of -4), and as a denominator, -1 (a divisor of 6), we get: $x = -\frac{2}{-1} = \frac{2}{1} = 2$.

Thus next, putting **2** into x in the polynomial P, we get:

$P = 6x^3 - 5x^2 - 12x - 4 = 6 \cdot 2^3 - 5 \cdot 2^2 - 12 \cdot 2 - 4 = 48 - 20 - 24 - 4 = 0$.

So we can see that $x - 2$ is a factor of P.

Thus next, applying to P the synthetic division by the factor $x - 2$, we get:

2	6	-5	-12	-4
		12	14	4
	6	7	2	0

So we get: $6x^3 - 5x^2 - 12x - 4 = (x - 2)(6x^2 + 7x + 2)$.

And in fact, $6x^2 + 7x + 2$ gets factorized to $(2x + 1)(3x + 2)$.

Thus, we get: $6x^3 - 5x^2 - 12x - 4 = (x - 2)(6x^2 + 7x + 2) = (x - 2)(2x + 1)(3x + 2)$.

In short:

Evaluating $6x^3 - 5x^2 - 12x - 4$ for $x = -\frac{2}{-1} = 2$, we get: $48 - 20 - 24 - 4 = 0$, so we get:

2	6	-5	-12	-4
		12	14	4
	6	7	2	0

Thus, we get: $6x^3 - 5x^2 - 12x - 4 = (x - 2)(6x^2 + 7x + 2) = (x - 2)(2x + 1)(3x + 2)$.

Examples F

Factorize the polynomials below.

0. $x^4 - 3x^3 - 15x^2 + 19x + 30$

1. $12x^4 + 46x^3 - 18x^2 - 36x + 16$

42

Suggestions or Solutions
To the **Problem** in the Example **0**

We have: $x^4 - 3x^3 - 15x^2 + 19x + 30$.

Factorizing it, we can get it the way below:

Evaluating $x^4 - 3x^3 - 15x^2 + 19x + 30$ for $x = -1$, we get: $1 + 3 - 15 - 19 + 30 = 0$, so doing the synthetic division, we get:

-1	1	-3	-15	19	30
		-1	4	11	-30
	1	-4	-11	30	0

So we get: $x^4 - 3x^3 - 15x^2 + 19x + 30 = (x + 1)(x^3 - 4x^2 - 11x + 30)$.

Next, evaluating $x^3 - 4x^2 - 11x + 30$ for $x = 2$, we get: $8 - 16 - 22 + 30 = 0$.

Thus, by the synthetic division, we get:

2	1	-4	-11	30
		2	-4	-30
	1	-2	-15	0

So we get: $x^3 - 4x^2 - 11x + 30 = (x - 2)(x^2 - 2x - 15) = (x - 2)(x + 3)(x - 5)$.

Therefore, $x^4 - 3x^3 - 15x^2 + 19x + 30 = (x + 1)(x - 2)(x + 3)(x - 5)$.

If not quite sure of the idea behind the processes above, follow the steps below:

Setting first, $P = x^4 - 3x^3 - 15x^2 + 19x + 30$, we can say that the polynomial P is of degree 4. So if P can get factorized, it will probably be a product of polynomials of degree 3 or less.

A polynomial of degree 3 can eventually get factorized to a product of polynomials of degree 1.   So eventually, **P** can get factorized to a product of polynomials of degree 1.

And thus, we can try factorizing **P** the way we factorize polynomials of degree 3 as

$x^3 + ux^2 + vx + w.$

So assuming that $x + e$ is a divisor of **P**, and setting:

$x^4 - 3x^3 - 15x^2 + 19x + 30 = (x + e)(ax^3 + bx^2 + cx + d)$, we get: $ed = 30$.

We can thus, assume that **e** and **d** are divisors of 30.   And we know:

$e^4 + 3e^3 - 15e^2 - 19e + 30 = (-e + e)(-ae^3 + be^2 - ce + d) = 0$ if $x = -e$, which is a divisor of 30, too, since **e** is a divisor of 30. And we know **e** is the negative of **-e**.

So putting a divisor of 30 into x in **P**, and getting **P = 0**, the negative of the divisor is the value of **e**.   By trial and error, we can try getting the divisor that makes **P = 0**.

Then, we get a factor of **P**. What is the factor though?

The factor is: $x + e$.   So dividing **P** by $x + e$, we get as the quotient a polynomial of degree 3 as $ax^3 + bx^2 + cx + d$.

Then, we can try factorizing it using the method described in the previous examples.

So let's now try factorizing $P = x^4 - 3x^3 - 15x^2 + 19x + 30$ using the idea above.

To begin with, taking -1 as a divisor of 30, and putting it into x in **P**, that is, evaluating

$x^4 - 3x^3 - 15x^2 + 19x + 30$ for $x = -1$, we get: $1 + 3 - 15 - 19 + 30 = 0$.

So we can see that $x + 1$ is a factor of P.

And thus, applying to P the synthetic division by the factor $x + 1$, we get:

$$
\begin{array}{c|ccccc}
-1 & 1 & -3 & -15 & 19 & 30 \\
   &   & -1 & 4 & 11 & -30 \\
\hline
   & 1 & -4 & -11 & 30 & 0
\end{array}
$$

So we get: $x^4 - 3x^3 - 15x^2 + 19x + 30 = (x + 1)(x^3 - 4x^2 - 11x + 30)$.

Next, we want to check to see if the quotient $x^3 - 4x^2 - 11x + 30$ can still be factorized.

Let's try this time, factorizing it a bit differently.

Assuming K is the quotient, and $(x + d)$ is a factor of K, we can set K this way too:

$K = x^3 - 4x^2 - 11x + 30 = (x + d)Q$, where Q is a polynomial of degree 2 as $ax^2 + bx + c$.

And actually setting: $x^3 - 4x^2 - 11x + 30 = (x + d)(ax^2 + bx + c)$, and expanding the right hand side, we can get: $dc = 30$.

What then, can we say about d and c?

We can assume that c and d are divisors of 30.   Then, $-c$ and $-d$ are divisors of 30, too.

And if we get: $K = 0$ when $x = -d$, we can say that $(x + d)$ is a factor of the polynomial K.

So putting a divisor of 30 into x in K, and getting: $K = 0$, we can say that the negative of the divisor can be the value of d.

Then, we can say that $(x + d)$ is a factor of K.

What if however, we get: $K = 0$ when $x = d$?

We have: $K = x^3 - 4x^2 - 11x + 30$. So we get: $K = d^3 - 4d^2 - 11d + 30$ if $x = d$.

So if $K = 0$ when $x = d$, that is, if $d^3 - 4d^2 - 11d + 30 = 0$, we can say that $x - d$ is a divisor of $x^3 - 4x^2 - 11x + 30$, and thus, is a factor of K. How come?

We can put K this way: too: $K = x^3 - 4x^2 - 11x + 30 = (x - d)(ax^2 + bx + c)$.

Then, we get: $K = 0$ when $x = d$. So $x - d$ divides K, and thus, is a factor of K.

So it does not matter whether we use as a factor $x + d$ or $x - d$.

If $K = 0$ when $x = d$, that is, if $d^3 - 4d^2 - 11d + 30 = 0$, we can say that $x - d$ is a divisor of $x^3 - 4x^2 - 11x + 30$, and thus, is a factor of K.

If $K = 0$ when $x = -d$, that is, if $-d^3 - 4d^2 + 11d + 30 = 0$, we can say that $x + d$ is a divisor of $x^3 - 4x^2 - 11x + 30$, and thus, is a factor of K.

And in fact, both of the two cases above are equivalent to each other.

- Let's now factorize $K = x^3 - 4x^2 - 11x + 30$ using the idea stated above.

To begin with, taking 2 as a divisor of 30, and putting it into x in K, that is, evaluating $x^3 - 4x^2 - 11x + 30$ for $x = 2$, we get: $2^3 - 4 \cdot 2^2 - 11 \cdot 2 + 30 = 8 - 16 - 22 + 30 = 0$.

So $x - 2$ is a factor of $x^3 - 4x^2 - 11x + 30$, and by the synthetic division, we get:

2	1	-4	-11	30
		2	-4	-30
	1	-2	-15	0

Thus, dividing $x^3 - 4x^2 - 11x + 30$ by $x - 2$, we get: $x^2 - 2x - 15$ as the quotient.

So we get: $x^3 - 4x^2 - 11x + 30 = (x - 2)(x^2 - 2x - 15)$.

Next, we want to check to see if the quotient $x^2 - 2x - 15$ can still be factorized.

Then, by the same token, we can set:

$x^2 - 2x - 15 = (x - c)Q$, where Q is a polynomials of degree 1 as $ax + b$.

So if $c^2 - 2c - 15 = 0$, $x - c$ is a divisor of $x^2 - 2x - 15$.

And in fact, we can get: $x^2 - 2x - 15 = 0$ for $x = -3$, which is a divisor of -15.

That is, evaluating $x^2 - 2x - 15$ for $x = -3$, we get: $(-3)^2 - 2 \cdot (-3) - 15 = 9 + 6 - 15 = 0$.

So $x + 3$ is a divisor, that is, a factor of $x^2 - 2x - 15$, and the synthetic division gives us:

```
-3 |  1      -2      -15
   |          -3       15
   |_____
      1      -5        0
```

Thus, dividing $x^2 - 2x - 15$ by $x + 3$, we get $x - 5$ as the quotient.

So we get: $x^2 - 2x - 15 = (x + 3)(x - 5)$.

Thus, putting threads together, we get:

$P = x^4 - 3x^3 - 15x^2 + 19x + 30 = (x + 1)(x^3 - 4x^2 - 11x + 30)$

$= (x + 1)(x - 2)(x^2 - 2x - 15) = (x + 1)(x - 2)(x + 3)(x - 5)$.

In short:

Evaluating $x^4 - 3x^3 - 15x^2 + 19x + 30$ for $x = -1$, we get: $1 + 3 - 15 - 19 + 30 = 0$, so doing the synthetic division, we get:

$$
\begin{array}{r|rrrrr}
-1 & 1 & -3 & -15 & 19 & 30 \\
   &   & -1 & 4   & 11 & -30 \\
\hline
   & 1 & -4 & -11 & 30 & 0
\end{array}
$$

So we get: $x^4 - 3x^3 - 15x^2 + 19x + 30 = (x + 1)(x^3 - 4x^2 - 11x + 30)$.

Next, evaluating $x^3 - 4x^2 - 11x + 30$ for $x = 2$, we get: $8 - 16 - 22 + 30 = 0$.

Thus, by the synthetic division, we get:

$$
\begin{array}{r|rrrr}
2 & 1 & -4 & -11 & 30 \\
  &   & 2 & -4 & -30 \\
\hline
  & 1 & -2 & -15 & 0
\end{array}
$$

So we get: $x^3 - 4x^2 - 11x + 30 = (x - 2)(x^2 - 2x - 15) = (x - 2)(x + 3)(x - 5)$.

Therefore, $x^4 - 3x^3 - 15x^2 + 19x + 30 = (x + 1)(x - 2)(x + 3)(x - 5)$.

Suggestions or Solutions
To the **Problem E** in the Example **0**

We have $12x^4 + 46x^3 - 18x^2 - 36x + 16$.

Factorizing it, we can get it the way below:

To begin with, $12x^4 + 46x^3 - 18x^2 - 36x + 16 = 2(6x^4 + 23x^3 - 9x^2 - 18x + 8)$.

So next, evaluating $6x^4 + 23x^3 - 9x^2 - 18x + 8$ for $x = \frac{1}{2}$, we get:

$6(\frac{1}{2})^4 + 23(\frac{1}{2})^3 - 9(\frac{1}{2})^2 - 18(\frac{1}{2}) + 8 = \frac{6}{2^4} + \frac{23}{2^3} - \frac{9}{2^2} - 9 + 8 = \frac{6+46-36}{2^4} - 1 = \frac{16}{2^4} - 1 = 0$. So we get:

$\frac{1}{2}$	6	23	-9	-18	8
		3	13	2	-8
	6	26	4	-16	0

So we get: $6x^4 + 23x^3 - 9x^2 - 18x + 8 = (x - \frac{1}{2})(6x^3 + 26x^2 + 4x - 16)$.

Next, evaluating $6x^3 + 26x^2 + 4x - 16$ for $x = \frac{2}{3}$, we get:

$6(\frac{2}{3})^3 + 26(\frac{2}{3})^2 + 4(\frac{2}{3}) - 16 = \frac{6\cdot8}{3^3} + \frac{26\cdot4}{3^2} + \frac{8}{3} - 16 = \frac{48+26\cdot4\cdot3+8\cdot9}{27} - 16 = \frac{48+26\cdot12+72}{27}$

$= \frac{12(4+26+6)}{27} - 16 = \frac{12\cdot36}{27} - 16 = \frac{12\cdot36-16\cdot27}{27} = \frac{4(3\cdot36-4\cdot27)}{27} = \frac{12(36-4\cdot9)}{27} = 0$. So we get:

$\frac{2}{3}$	6	26	4	-16
		4	20	16
	6	30	24	0

So we get: $6x^3 + 26x^2 + 4x - 16 = (x - \frac{2}{3})(6x^2 + 30x + 24) = 6(x - \frac{2}{3})(x^2 + 5x + 4)$

$= 6(x - \frac{2}{3})(x + 1)(x + 4)$.

Thus, $12x^4 + 46x^3 - 18x^2 - 36x + 16 = 2(x - \frac{1}{2})6(x - \frac{2}{3})(x + 1)(x + 4)$

$= (2x - 1)2(3x - 2)(x + 1)(x + 4) = 2(2x - 1)(3x - 2)(x + 1)(x + 4)$.

If not quite sure of the idea behind the processes above, follow the steps below:

Let's set first, $P = 12x^4 + 46x^3 - 18x^2 - 36x + 16$.

To begin with, 2 is a divisor common to all the terms in P, so 2 is a factor of P.

Thus, we get: $P = 2(6x^4 + 23x^3 - 9x^2 - 18x + 8)$.

So next, we want to get $6x^4 + 23x^3 - 9x^2 - 18x + 8$ factorized.

We know a polynomial $sx^3 + tx^2 + ux + v$ can get factorized to $(ax + b)(cx + d)(ex + g)$.

So we can reasonably expect that a polynomial $sx^4 + tx^3 + ux^2 + vx + w$ can eventually get factorized to $(ax + b)(cx + d)(ex + g)(hx + k)$.

And thus, setting: $sx^4 + tx^3 + ux^2 + vx + w = (ax + b)(cx + d)(ex + g)(hx + k)$, and expanding the right hand side, we get a polynomial as below:

$acehx^4 + (acek + \text{etc.})x^3 + (acgk + \text{etc.})x^2 + (adgk + \text{etc.})x + bdgk$.

Then, we can see that $s = aceh$, and that $w = bdgk$.

So assuming that a, c, e, and h are divisors of s, and b, d, g, and k are divisors of w, and finding a, b, c, d, e, g, h, and k by the similar method applied to $x^4 + ux^3 + vx^2 + wx + s$, we could get the factorization of $sx^4 + tx^3 + ux^2 + vx + w$.

How can we find a, b, c, d, e, g, h, and k, though?

Setting $A = sx^4 + tx^3 + ux^2 + vx + w = (ax + b)(cx + d)(ex + g)(hx + k)$, we can see that:

$A = 0$ when $x = -\frac{b}{a}, -\frac{d}{c}, -\frac{g}{e}$, or $-\frac{k}{h}$.

Thus, if $m = \frac{b}{a}, \frac{d}{c}, \frac{g}{e}$, or $\frac{k}{h}$, then $x + m$ is a factor of A.

And we can put it the way below, too:

If $m = -\frac{b}{a}, -\frac{d}{c}, -\frac{g}{e}$, or $-\frac{k}{h}$, then $x - m$ is a factor of A.

So first, we can assume that a, c, e, and h are divisors of s, and b, d, g, and k are divisors of w.

And by trial and error, coming up with, for instance, $-\frac{b}{a}$ that makes $A = 0$, we can get a factor of A. And the factor is: $x + \frac{b}{a}$.

And next, we can get the other factors factorizing the quotient we get after applying to A the synthetic division by the factor.

And of course, if the quotient is prime, we can use the quotient itself as the other factor.

So let's now, set: $Q = 6x^4 + 23x^3 - 9x^2 - 18x + 8$, and factorize Q using the idea above.

Then, first, we want to find a value of x that makes $Q = 0$. And the value is a fraction where the numerator is a divisor of 8, and the denominator is a divisor of 6.

(The divisors of 8 are ± 1, ± 2, ± 4, and ± 8, and the divisors of 6 are ± 1, ± 2, ± 3, and ± 6.)

So for instance, using as a numerator, 1 (a divisor of 8), and as a denominator, -2 (a divisor of 6), we get: $x = -\frac{1}{-2} = \frac{1}{2}$.

Thus next, putting $\frac{1}{2}$ into x in the polynomial Q, we get:

$Q = 6(\frac{1}{2})^4 + 23(\frac{1}{2})^3 - 9(\frac{1}{2})^2 - 18(\frac{1}{2}) + 8 = \frac{6}{2^4} + \frac{23}{2^3} - \frac{9}{2^2} - 9 + 8 = \frac{6+46-36}{2^4} - 1 = \frac{16}{2^4} - 1 = 0$.

So we can see that $x - \frac{1}{2}$ is a factor of Q. Thus next, applying to Q the synthetic division by the factor $x - \frac{1}{2}$, we get:

$\frac{1}{2}$	6	23	-9	-18	8
		3	13	2	-8
	6	26	4	-16	0

So we get: $Q = 6x^4 + 23x^3 - 9x^2 - 18x + 8 = (x - \frac{1}{2})(6x^3 + 26x^2 + 4x - 16)$.

Next, we want to check to see if $6x^3 + 26x^2 + 4x - 16$ can be factorized.

Suppose $T = 6x^3 + 26x^2 + 4x - 16$.

Then, by the same token, first, we want to find a value of x that makes $T = 0$. And the value is a fraction where the numerator is a divisor of -16, and the denominator is a divisor of 6.

(The divisors of -16 are ± 1, ± 2, ± 4, ± 8, and ± 16, and those of 6 are ± 1, ± 2, ± 3, and ± 6.)

So for instance, using as a numerator, 2 (a divisor of -16), and as a denominator, -3 (a divisor of 6), we get: $x = -\frac{2}{-3} = \frac{2}{3}$.

Thus next, putting $\frac{2}{3}$ into x in the polynomial T, we get:

$$T = 6(\tfrac{2}{3})^3 + 26(\tfrac{2}{3})^2 + 4(\tfrac{2}{3}) - 16 = \tfrac{6 \cdot 8}{3^3} + \tfrac{26 \cdot 4}{3^2} + \tfrac{8}{3} - 16 = \tfrac{48 + 26 \cdot 4 \cdot 3 + 8 \cdot 9}{27} - 16 = \tfrac{48 + 26 \cdot 12 + 72}{27}$$

$$= \tfrac{12(4 + 26 + 6)}{27} - 16 = \tfrac{12 \cdot 36}{27} - 16 = \tfrac{12 \cdot 36 - 16 \cdot 27}{27} = \tfrac{4(3 \cdot 36 - 4 \cdot 27)}{27} = \tfrac{12(36 - 4 \cdot 9)}{27} = 0.$$

So we can see that $x - \frac{2}{3}$ is a factor of T. Thus, doing the synthetic division, we get:

$\frac{2}{3}$	6	26	4	-16
		4	20	16
	6	30	24	0

So we get: $6x^3 + 26x^2 + 4x - 16 = (x - \tfrac{2}{3})(6x^2 + 30x + 24) = 6(x - \tfrac{2}{3})(x^2 + 5x + 4)$

$= 6(x - \tfrac{2}{3})(x + 1)(x + 4)$

Now, putting threads together, we get:

$$P = 2(6x^4 + 23x^3 - 9x^2 - 18x + 8) = 2(x - \tfrac{1}{2})(6x^3 + 26x^2 + 4x - 16)$$

$$= 2(x - \tfrac{1}{2})6(x - \tfrac{2}{3})(x + 1)(x + 4) = (2x - 1)2(3x - 2)(x + 1)(x + 4).$$

So $12x^4 + 46x^3 - 18x^2 - 36x + 16$ gets factorized to $2(2x - 1)(3x - 2)(x + 1)(x + 4)$.

And we can try factorizing $Q = 6x^4 + 23x^3 - 9x^2 - 18x + 8$ the way below, too:

Assuming first, $x + e$ is a divisor of Q, and setting:

$6x^4 + 23x^3 - 9x^2 - 18x + 8 = (x + e)(ax^3 + bx^2 + cx + d)$, we get: $ed = 8$.

Thus next, we can assume that e and d are divisors of 8. And we know:

$6e^4 + 23e^3 - 9e^2 - 18e + 8 = (-e + e)(-ae^3 + be^2 - ce + d) = 0$ if $x = -e$, which is a divisor of 8, too, since e is a divisor of 8. And we know e is the negative of $-e$.

So putting a divisor of 8 into x in Q, and getting $Q = 0$, we can say that the negative of the divisor is the value of e.

By trial and error, we can try getting the divisor that makes $Q = 0$.

Then, we get a factor of Q. And the factor is: $x + e$. So dividing Q by $x + e$, we get as the quotient a polynomial of degree 3 as $ax^3 + bx^2 + cx + d$.

Then, we can try factorizing it using the method described in the previous examples.

In short:

To begin with, $12x^4 + 46x^3 - 18x^2 - 36x + 16 = 2(6x^4 + 23x^3 - 9x^2 - 18x + 8)$.

So next, evaluating $6x^4 + 23x^3 - 9x^2 - 18x + 8$ for $x = \frac{1}{2}$, we get:

$6(\frac{1}{2})^4 + 23(\frac{1}{2})^3 - 9(\frac{1}{2})^2 - 18(\frac{1}{2}) + 8 = \frac{6}{2^4} + \frac{23}{2^3} - \frac{9}{2^2} - 9 + 8 = \frac{6 + 46 - 36}{2^4} - 1 = \frac{16}{2^4} - 1 = 0$. So we get:

$\frac{1}{2}$	6	23	-9	-18	8
		3	13	2	-8
	6	26	4	-16	0

So we get: $6x^4 + 23x^3 - 9x^2 - 18x + 8 = (x - \frac{1}{2})(6x^3 + 26x^2 + 4x - 16)$.

Next, evaluating $6x^3 + 26x^2 + 4x - 16$ for $x = \frac{2}{3}$, we get:

$6(\frac{2}{3})^3 + 26(\frac{2}{3})^2 + 4(\frac{2}{3}) - 16 = \frac{6 \cdot 8}{3^3} + \frac{26 \cdot 4}{3^2} + \frac{8}{3} - 16 = \frac{48 + 26 \cdot 4 \cdot 3 + 8 \cdot 9}{27} - 16 = \frac{48 + 26 \cdot 12 + 72}{27}$

$= \frac{12(4 + 26 + 6)}{27} - 16 = \frac{12 \cdot 36}{27} - 16 = \frac{12 \cdot 36 - 16 \cdot 27}{27} = \frac{4(3 \cdot 36 - 4 \cdot 27)}{27} = \frac{12(36 - 4 \cdot 9)}{27} = 0.$

So we get:

$\frac{2}{3}$	6	26	4	-16
		4	20	16
	6	30	24	0

So we get: $6x^3 + 26x^2 + 4x - 16 = (x - \frac{2}{3})(6x^2 + 30x + 24) = 6(x - \frac{2}{3})(x^2 + 5x + 4)$

$= 6(x - \frac{2}{3})(x + 1)(x + 4)$.

Thus, $12x^4 + 46x^3 - 18x^2 - 36x + 16 = 2(x - \frac{1}{2})6(x - \frac{2}{3})(x + 1)(x + 4)$

$= (2x - 1)2(3x - 2)(x + 1)(x + 4) = 2(2x - 1)(3x - 2)(x + 1)(x + 4)$.

Examples G

0.   Factorize the polynomial below:

$$x^3 + 2x^2y + xy^2 + 4xy + 2x^2z + xz^2 + 4xz + 2xyz + 2x^2 + 2y^2 + 2z^2 + 4yz$$

1.   Find the value of $a^2 + b^2 + c^2$ in each case below:

1.0. $a + b + c = 25$, and $ab + bc + ca = 37$.

1.1. $a - b - c = 32$, and $bc - ab - ca = 82$.

1.2. $c - a - b = 27$, and $ab - bc - ca = 98$.

1.3. $a - b - c = 12$, and $ab - bc + ca = 21$.

Suggestions or Solutions
To the **Problem** in the Example **0**

We have: $x^3 + 2x^2y + xy^2 + 4xy + 2x^2z + xz^2 + 4xz + 2xyz + 2x^2 + 2y^2 + 2z^2 + 4yz$.

Factorizing it, we can get it the way below:

$$P = x^3 + 2x^2y + xy^2 + 4xy + 2x^2z + xz^2 + 4xz + 2xyz + 2x^2 + 2y^2 + 2z^2 + 4yz$$

$$= x^3 + 2(y + z + 1)x^2 + (y^2 + z^2 + 2yz + 4y + 4z)x + 2(y^2 + z^2 + 2yz)$$

$$= x^3 + 2(y + z + 1)x^2 + \{(y + z)^2 + 4(y + z)\}x + 2(y + z)^2$$

$$= x^3 + 2(y + z + 1)x^2 + (y + z)(y + z + 4)x + 2(y + z)^2$$

So setting $t = y + z$, we get: $P = x^3 + 2(t + 1)x^2 + t(t + 4)x + 2t^2$.

Next, evaluating P when $x = -2$, we get:

$$(-2)^3 + 2(t + 1)(-2)^2 + t(t + 4)(-2) + 2t^2 = -8 + 8t + 8 - 2t^2 - 8t + 2t^2 = 0.$$

So $x + 2$ is a factor, and thus, doing the synthetic division, we get:

-2	1	$2(t + 1)$	$t(t + 4)$	$2t^2$
		-2	-4t	$-2t^2$
	1	$2t$	t^2	0

Thus, we get: $P = (x + 2)(x^2 + 2tx + t^2) = (x + 2)(x + t)^2$.

We know: $t = y + z$. So we get: $P = (x + 2)(x + y + z)^2$.

If not quite sure of the idea behind the processes above, follow the steps below:

Suppose the polynomials given is P. Though the polynomial P looks quite complicated, yet factorizing it, we will get to see it's fairly simple.

Examining **P**, we can notice that we can take **P** for a polynomial of degree 3 with respect to **x**.   So first, we may want to put **P** in terms of **x**.   Then, we get:

$$P = x^3 + 2x^2y + xy^2 + 4xy + 2x^2z + xz^2 + 4xz + 2xyz + 2x^2 + 2y^2 + 2z^2 + 4yz$$

$$= x^3 + 2x^2y + 2x^2z + 2x^2 + xy^2 + xz^2 + 2xyz + 4xy + 4xz + 2y^2 + 2z^2 + 4yz$$

$$= x^3 + 2x^2(y + z + 1) + x(y^2 + z^2 + 2yz + 4y + 4z) + 2y^2 + 2z^2 + 4yz$$

$$= x^3 + 2(y + z + 1)x^2 + (y^2 + z^2 + 2yz + 4y + 4z)x + 2(y^2 + z^2 + 2yz).$$

Now, setting $u = 2(y + z + 1)$, $v = y^2 + z^2 + 2yz + 4y + 4z$, and $w = 2(y^2 + z^2 + 2yz)$, we get: $P = x^3 + ux^2 + vx + w$.

So we can take **P** for a polynomial of degree 3 with respect to **x**, and therefore, we can consider factorizing **P** the way we do to the polynomial $x^3 + ux^2 + vx + w$, which can be factorized to $(x + a)(x + b)(x + c)$.

Let's next, look at the coefficients and the constant term in $P = x^3 + ux^2 + vx + w$.

Then first, looking closely at $w = 2(y^2 + z^2 + 2yz)$, we can notice that $w = 2(y + z)^2$.

Moving next, on to $v = y^2 + z^2 + 2yz + 4y + 4z$, we can notice that it can be put this way, too: $y^2 + z^2 + 2yz + 4y + 4z = (y + z)^2 + 4(y + z) = (y + z)\{(y + z) + 4\} = (y + z)(y + z + 4)$.

So we get: $v = (y + z)(y + z + 4)$.

Thus, we now have: $P = x^3 + 2(y + z + 1)x^2 + (y + z)(y + z + 4)x + 2(y + z)^2$, which looks a bit simpler, and is still in the form of $x^3 + ux^2 + vx + w$.

What about the coefficient $u = 2(y + z + 1)$, though?

We can notice that $y + z$ is common to the two coefficients **u** and **v**, and the constant term **w**. That is, it is common to $2(y + z + 1)$, $(y + z)(y + z + 4)$, and $2(y + z)^2$.

So we may want to set it to another constant t, for instance.

Thus, setting $t = y + z$, we get: $P = x^3 + 2(t + 1)x^2 + t(t + 4)x + 2t^2$, which now looks much simpler.

Also of course, P is still in the form of $x^3 + ux^2 + vx + w$.

Thus, we can readily apply to P above the method we use factorizing polynomials of the form $x^3 + ux^2 + vx + w$, which we know, can be factorized to $(x + a)(x + b)(x + c)$.

So first, we want to collect the divisors of $2t^2$, and then, for instance, find a divisor called d that makes $x - d$ a factor of P.

The divisors of $2t^2$ are ± 1, ± 2, $\pm t$, $2t$, $\pm t^2$, and $\pm 2t^2$.

So let's first check to see if 1 can work. Then, putting 1 into x, we get:

$1 + 2(t + 1) + t(t + 4) + 2t^2 = 3t^2 + 6t + 3 = 3(t^2 + 2t + 1) = 3(t + 1)^2$, which can be 0 when $t = -1$ only, and thus, is not 0 for every value of t.

If $x - 1$ is a factor of P, P has to be 0 for all values of t when $x = 1$. So 1 doesn't work.

So let's next, see if -2 works. Then, we get:
$(-2)^3 + 2(t + 1)(-2)^2 + t(t + 4)(-2) + 2t^2 = -8 + 8t + 8 - 2t^2 - 8t + 2t^2 = 0$.

So we get: $P = 0$ for all values of t when $x = -2$, and thus, -2 works.

So $x + 2$ is a factor of $P = x^3 + 2(t + 1)x^2 + t(t + 4)x + 2t^2$.

Thus, doing the synthetic division, we get:

$$
\begin{array}{r|cccc}
-2 & 1 & 2(t + 1) & t(t + 4) & 2t^2 \\
   &   & -2 & -4t & -2t^2 \\
\hline
   & 1 & 2t & t^2 & 0
\end{array}
$$

So dividing $x^3 + 2(t + 1)x^2 + t(t + 4)x + 2t^2$ by $x + 2$, we get $x^2 + 2tx + t^2$ as the quotient.

Thus, we get: $P = x^3 + 2(t + 1)x^2 + t(t + 4)x + 2t^2 = (x + 2)(x^2 + 2tx + t^2)$.

Next, the quotient $x^2 + 2tx + t^2$ can be factorized to $(x + t)^2$.

So we get: $P = x^3 + 2(t + 1)x^2 + t(t + 4)x + 2t^2 = (x + 2)(x + t)^2$.

We know: $t = y + z$. So we get: $P = (x + 2)(x + y + z)^2$.

In short:

$P = x^3 + 2x^2y + xy^2 + 4xy + 2x^2z + xz^2 + 4xz + 2xyz + 2x^2 + 2y^2 + 2z^2 + 4yz$

$= x^3 + 2(y + z + 1)x^2 + (y^2 + z^2 + 2yz + 4y + 4z)x + 2(y^2 + z^2 + 2yz)$

$= x^3 + 2(y + z + 1)x^2 + \{(y + z)^2 + 4(y + z)\}x + 2(y + z)^2$

$= x^3 + 2(y + z + 1)x^2 + (y + z)(y + z + 4)x + 2(y + z)^2$

So setting $t = y + z$, we get: $P = x^3 + 2(t + 1)x^2 + t(t + 4)x + 2t^2$.

Next, evaluating P when $x = -2$, we get:

$(-2)^3 + 2(t + 1)(-2)^2 + t(t + 4)(-2) + 2t^2 = -8 + 8t + 8 - 2t^2 - 8t + 2t^2 = 0$.

So $x + 2$ is a factor, and thus, doing the synthetic division, we get:

-2	1	$2(t + 1)$	$t(t + 4)$	$2t^2$
		-2	$-4t$	$-2t^2$
	1	$2t$	t^2	0

Thus, we get: $P = (x + 2)(x^2 + 2tx + t^2) = (x + 2)(x + t)^2$.

We know: $t = y + z$. So we get: $P = (x + 2)(x + y + z)^2$.

Suggestions or Solutions
To the **Problems** in the Example **1**

Find the value of $a^2 + b^2 + c^2$ in each case below:

1.0. $a + b + c = 25$, and $ab + bc + ca = 37$.

1.1. $a - b - c = 32$, and $bc - ab - ca = 82$.

1.2. $c - a - b = 27$, and $ab - bc - ca = 98$.

1.3. $a - b - c = 12$, and $ab - bc + ca = 21$.

1.0.

We have a factorization identity where $(a + b + c)^2 = a^2 + b^2 + c^2 + 2(ab + bc + ca)$.

So setting $A = a^2 + b^2 + c^2$, since $a + b + c = 25$, and $ab + bc + ca = 37$, we get:

$25^2 = A + 2 \cdot 37 \Rightarrow A = 625 - 74 = 551 \Rightarrow A = 551$.

1.1.

This is just another algebra practice.

In the identity: $(a + b + c)^2 = a^2 + b^2 + c^2 + 2(ab + bc + ca)$, we can notice that replacing b with $-b$, and c with $-c$, we get:

$(a - b - c)^2 = a^2 + b^2 + c^2 + 2\{-ab + (-b)(-c) - ca\} = a^2 + b^2 + c^2 + 2(bc - ab - ca)$.

Thus, we get: $(a - b - c)^2 = a^2 + b^2 + c^2 + 2(bc - ab - ca)$.

So setting $A = a^2 + b^2 + c^2$, since $a - b - c = 32$, and $bc - ab - ca = 82$, we get:

$32^2 = A + 2 \cdot 82 \Rightarrow A = 32^2 - 2 \cdot 82 = (30 + 2)32 - 2 \cdot 82 = 30 \cdot 32 + 2 \cdot 32 - 2 \cdot 82$

$= 30 \cdot 32 + 2(32 - 82) = 960 + 2(-50) = 960 - 100 = 860 \Rightarrow A = 860$.

1.2.   Find the value of $a^2 + b^2 + c^2$ in the case below:

$c - a - b = 27$, and $ab - bc - ca = 98$.

We have a factorization identity where $(a + b + c)^2 = a^2 + b^2 + c^2 + 2(ab + bc + ca)$.

We can notice that in the identity above, replacing *a* with *-a*, and *b* with *-b*, we get:

$(-a - b + c)^2 = a^2 + b^2 + c^2 + 2\{(-a)(-b) - bc - ca\} = a^2 + b^2 + c^2 + 2(ab - bc - ca)$.

Thus, we get: $(c - a - b)^2 = a^2 + b^2 + c^2 + 2(ab - bc - ca)$.

So setting $A = a^2 + b^2 + c^2$, since $c - a - b = 27$, and $ab - bc - ca = 98$, we get:

$32^2 = A + 2 \cdot 82 \Rightarrow A = 27^2 - 2 \cdot 98 = (20 + 7)27 - 2 \cdot 98 = 20 \cdot 27 + 7 \cdot 27 - 2 \cdot (100 - 2)$

$= 20(20 + 7) + 7(20 + 7) - 200 + 4 = 400 + 140 + 140 + 49 - 196$

$= 540 + 49 - 59 = 540 - 10 = 530 \Rightarrow A = 530$.

1.3. **Find the value of $a^2 + b^2 + c^2$ in the case below:**

$a - b - c = 12$, **and** $ab - bc + ca = 21$.

Again, in identity: $(a + b + c)^2 = a^2 + b^2 + c^2 + 2(ab + bc + ca)$, we can notice that replacing **b** with **-b**, and **c** with **-c**, we get:

$(a - b - c)^2 = a^2 + b^2 + c^2 + 2\{-ab + (-b)(-c) - ca\} = a^2 + b^2 + c^2 + 2(bc - ab - ca)$.

Thus, we get: $(a - b - c)^2 = a^2 + b^2 + c^2 + 2(bc - ab - ca)$.

However, we are given the value of **ab − bc + ca** and not **bc − ab − ca**.
So what can we do?

We can notice that **bc − ab − ca** = **-(ab − bc + ca)** = **-21**.

So setting $A = a^2 + b^2 + c^2$, since we have: $(a - b - c)^2 = a^2 + b^2 + c^2 + 2(bc - ab - ca)$, together with **a − b − c = 12**, and **bc − ab − ca = -21**, we get:

$12^2 = A + 2 \cdot (-21)$

$\Rightarrow A = 12^2 + 2 \cdot 21 = (10 + 2)12 - 42 = 120 + 24 - 42 = 144 - 42 = 102$

$\Rightarrow A = 102$.

Examples H

Factorize each of the following polynomials.

0. $a^3 + b^3$

1. $a^3 - b^3$

Suggestions or Solutions
To the **Problem** in the Example **0**

We have: $a^3 + b^3$.

Factorizing it, we can get it the way below:

$$a^3 + b^3 = a^3 + a^2b - a^2b - ab^2 + ab^2 + b^3 = a^2(a + b) - a^2b - ab^2 + b^2(a + b)$$

$$= (a + b)(a^2 + b^2) - a^2b - ab^2 = (a + b)(a^2 + b^2) - ab(a + b) = (a + b)(a^2 - ab + b^2).$$

If not quite sure of the idea behind the processes above, follow the steps below:

Factorizing an integer or a polynomial, we are expecting that it has divisors. And such divisors are called factors.

So factorizing a polynomial, we are expecting that the polynomial has factors.

And if factors exist, the polynomial has two factors at least, which can be the same, though.   How come?

For instance, x is a factor of a polynomial $x^2 + 3x$, so we get: $x^2 + 3x = x(x + 3)$. Thus, the polynomial has two factors, which are x and $x + 3$.

For another instance, x is a factor of $x^3 + 3x^2 + 2x$, so we get:

$x^3 + 3x^2 + 2x = x(x^2 + 3x + 2)$, which in turn, gets factorized to $x(x + 1)(x + 2)$.

For one more instance, $x^2 + 2x + 1$ gets factorized to $(x + 1)^2$, and therefore, has as factors two of $(x + 1)$s.

And a factor can have its factors, too.

For instance, $x(x^2 + 3x + 2)$ has $x^2 + 3x + 2$ as a factor, which has two factors, which are $x + 1$ and $x + 2$, because: $x^2 + 3x + 2 = (x + 1)(x + 2)$.

And if a factor is prime, that is, if it is a prime factor, it is often said to have no factor.

So just saying that an integer or an expression has no factor, we mean it has as divisors 1 and itself only.

And thus, if a factor has no divisor other than 1 and itself, it is often said to have no factor, and is said to be prime, so it is called a prime factor.

Now, for instance, the polynomial $x^2 + 2x + 1 = (x + 1)^2$ has two same prime factors, because $x + 1$ is prime, and the polynomial $x^3 + 3x^2 + 2x = x(x + 1)(x + 2)$ has three prime factors, which are $x, x + 1$, and $x + 2$.

So factorizing a polynomial, we are expecting the polynomial to have two or more prime factors. And such a factor can be a polynomial, a monomial, or a constant as well as an integer, of course.   So what?

Once having found all the prime factors, we put all those factors in a form of a product. Then, the form is called the factorization.

Therefore, for instance, taking a product of polynomials, that is, expanding the product, we get a polynomial factorable, which sounds, of course, quite natural though.

How then, can we factorize the polynomial $a^3 + b^3$ given in this example?

Let's set first, $P = a^3 + b^3$.
The polynomial P has two terms, which however, have nothing in common.

In other words, there is no divisor common to both of the terms. So is P not factorable?

Even having no divisor common to all the terms, it can still be factorized.

We know multiplying two polynomials, we get a polynomial that is factorable.

Multiplying for instance, $x + 1$ by $x + 2$, we get: $x^2 + 3x + 2$, which is thus, factorable.

So if P is factorable, there must be two polynomials that can get multiplied together to produce the two terms, a^3 and b^3. Thus, we may want to think of such two polynomials.

For instance, we can start with two polynomials $a + b$ and $a^2 + b^2$.

Then, expanding $(a + b)(a^2 + b^2)$, we get: $(a + b)(a^2 + b^2) = a^3 + ab^2 + a^2b + b^3$.

So we now have $a^3 + b^3$, but we don't want $ab^2 + a^2b$.   What do we do then?

We can try doing something to the product $(a + b)(a^2 + b^2)$ so that the two unwanted terms get canceled during the expansion.

That is, we want $ab^2 + a^2b$ to vanish. It doesn't vanish itself, though, of course.

So in other words, we want $-(ab^2 + a^2b)$ to be made, too, during the expansion.   How?

Looking at $-(ab^2 + a^2b)$ a bit more closely, we can notice that there is a common divisor, and it is ab. That is, $-(ab^2 + a^2b)$ can be factorized to $-ab(a + b)$.

What then, can we do?

Looking at **-ab(a + b)** a little more closely, we can notice that it is the product of **-ab** and **(a + b)**.   So what?

We can add **-ab** to the polynomial $a^2 + b^2$.   Then, what?

We know **(a + b)(a² + b²)** produces $a^3 + ab^2 + a^2b + b^3$, which is $a^3 + ab(a + b) + b^3$.

So **(a + b){a² + b² + (-ab)}** produces not only $a^3 + ab(a + b) + b^3$ but -ab(a + b), too. So?

So adding **-ab** to the polynomial $a^2 + b^2$ in the product **(a + b)(a² + b²)**, we get:

$(a + b)(a^2 + b^2 - ab) = a^3 + ab(a + b) + b^3 - ab(a + b) = a^3 + b^3$.

Now, each of **(a + b)** and **(a² + b² – ab)** is prime.

Therefore, $a^3 + b^3$ gets factorized to **(a + b)(a² + b² – ab)**.

In short:

$$a^3 + b^3 = a^3 + a^2b - a^2b - ab^2 + ab^2 + b^3 = a^2(a + b) - a^2b - ab^2 + b^2(a + b)$$

$$= (a + b)(a^2 + b^2) - a^2b - ab^2 = (a + b)(a^2 + b^2) - ab(a + b) = (a + b)(a^2 - ab + b^2).$$

Doing arithmetic, we sometimes subtract values from other values.
Running math though, we always compensate the values, that is, add the negatives.

Suggestions or Solutions
To the **Problem** in the Example **1**

We have: $a^3 - b^3$.

Let's set first, $P = a^3 - b^3$.
In the Example 0 above, we have: $a^3 + b^3 = (a + b)(a^2 - ab + b^2)$.

Replacing b with $-b$ in $a^3 + b^3$, we get: $a^3 + (-b)^3 = a^3 - b^3$, which is the polynomial P.

So we can simply get: $P = a^3 - b^3 = \{a + (-b)\}\{a^2 - a(-b) + (-b)^2\} = (a - b)(a^2 + ab + b^2)$.

Thus, we get: $P = a^3 - b^3 = (a - b)(a^2 + ab + b^2)$.

So putting threads together, we get: $a^3 \pm b^3 = (a \pm b)(a^2 \mp ab + b^2)$.

In short:

$a^3 - b^3 = a^3 - a^2b + a^2b - ab^2 + ab^2 - b^3 = a^2(a - b) + a^2b - ab^2 + b^2(a - b)$

$= (a - b)(a^2 + b^2) + a^2b - ab^2 = (a - b)(a^2 + b^2) + ab(a - b) = (a - b)(a^2 + ab + b^2)$.

How about the factorizations of polynomials as $a^4 \pm b^4$ and $a^5 \pm b^5$?

(The material starts below and ends before the next example is for those students who want to go beyond the basics.)

We have a factorization identity where $(x + y)^2 = x^2 + 2xy + y^2$.
So replacing x with a^2, and y with b^2, we get: $(a^2 + b^2)^2 = a^4 + 2a^2b^2 + b^4$.
Thus, we get: $a^4 + b^4 = (a^2 + b^2)^2 - 2a^2b^2$.
Also, we have another factorization identity where $x^2 - y^2 = (x + y)(x - y)$.

So we get: $a^4 + b^4 = (a^2 + b^2)^2 - 2a^2b^2 = \{(a^2 + b^2) + \sqrt{2}ab\}\{(a^2 + b^2) - \sqrt{2}ab\}$

$= (a^2 + b^2 + \sqrt{2}ab)(a^2 + b^2 - \sqrt{2}ab)$.

Thus, we get: $a^4 + b^4 = (a^2 + b^2 + \sqrt{2}ab)(a^2 + b^2 - \sqrt{2}ab)$.

Next, using the identity $x^2 - y^2 = (x + y)(x - y)$ twice, we get:

$$a^4 - b^4 = (a^2 + b^2)(a^2 - b^2) = (a^2 + b^2)(a + b)(a - b).$$

So we get: $a^4 - b^4 = (a^2 + b^2)(a + b)(a - b)$.

Next, taking the product of $a + b$ and $a^4 + b^4$, we can get $a^5 + b^5$, along with other terms.

So taking care of the other terms, we can get the factorization of $a^5 + b^5$.

Now, taking the product first, we get:

$$(a + b)(a^4 + b^4) = a^5 + ab^4 + a^4b + b^5 = a^5 + b^5 + ab(a^3 + b^3).$$

And we have a factorization identity where $x^3 \pm y^3 = (x \pm y)(x^2 \mp xy + y^2)$. So we get:

$$(a + b)(a^4 + b^4) = a^5 + b^5 + ab(a + b)(a^2 - ab + b^2) = a^5 + b^5 + (a + b)ab(a^2 - ab + b^2)$$
$$= a^5 + b^5 + (a + b)(a^3b - a^2b^2 + ab^3). \quad \text{Thus, we get:}$$

$$a^5 + b^5 = (a + b)(a^4 + b^4) - (a + b)(a^3b - a^2b^2 + ab^3)$$
$$= (a + b)\{(a^4 + b^4) - (a^3b - a^2b^2 + ab^3)\} = (a + b)(a^4 + b^4 - a^3b + a^2b^2 - ab^3).$$

Therefore, we get: $a^5 + b^5 = (a + b)(a^4 - a^3b + a^2b^2 - ab^3 + b^4)$.

Also, replacing b with $-b$ in $a^5 + b^5$, we get: $a^5 - b^5$, so we get:

$$a^5 - b^5 = (a - b)(a^4 + a^3b + a^2b^2 + ab^3 + b^4).$$

Thus, putting threads together, we get: $a^5 \pm b^5 = (a \pm b)(a^4 \mp a^3b + a^2b^2 \mp ab^3 + b^4)$.

How about the factorizations of polynomials as $a^6 \pm b^6$ and $a^7 \pm b^7$?

We can use again the factorization identity where $(x + y)^2 = x^2 + 2xy + y^2$.

So replacing x with a^3, and y with b^3, we get: $(a^3 + b^3)^2 = a^6 + 2a^3b^3 + b^6$.

Thus, we get: $a^6 + b^6 = (a^3 + b^3)^2 - 2a^3b^3$.

Also, we have another factorization identity where $x^2 - y^2 = (x + y)(x - y)$.

So we get: $a^6 + b^6 = (a^3 + b^3)^2 - 2a^3b^3 = \{(a^3 + b^3) + \sqrt{2}a^{\frac{3}{2}}b^{\frac{3}{2}}\}\{(a^2 + b^2) - \sqrt{2}a^{\frac{3}{2}}b^{\frac{3}{2}}\}$

$= (a^3 + b^3 + \sqrt{2}a^{\frac{3}{2}}b^{\frac{3}{2}})(a^2 + b^2 - \sqrt{2}a^{\frac{3}{2}}b^{\frac{3}{2}})$.

Thus, we get: $a^6 + b^6 = (a^3 + b^3 + \sqrt{2}a^{\frac{3}{2}}b^{\frac{3}{2}})(a^2 + b^2 - \sqrt{2}a^{\frac{3}{2}}b^{\frac{3}{2}})$.

Next, using the identities $x^2 - y^2 = (x + y)(x - y)$ and $x^3 - y^3 = (x - y)(x^2 + xy + y^2)$, we get: $a^6 - b^6 = (a^3 + b^3)(a^3 - b^3) = (a^3 + b^3)(a - b)(a^2 + ab + b^2)$.

So we get: $a^6 - b^6 = (a^3 + b^3)(a - b)(a^2 + ab + b^2)$.

Next, taking the product of $a + b$ and $a^6 + b^6$, we can get: $a^7 + b^7$, along with other terms.

So removing the other terms, we can get the factorization of $a^7 + b^7$.

Now, taking the product first, we get:
$(a + b)(a^6 + b^6) = a^7 + ab^6 + a^6b + b^7 = a^7 + b^7 + ab(a^5 + b^5)$.

We have a factorization identity where $a^5 \pm b^5 = (a \pm b)(a^4 \mp a^3b + a^2b^2 \mp ab^3 + b^4)$.

So we get: $(a + b)(a^6 + b^6) = a^7 + b^7 + ab(a + b)(a^4 - a^3b + a^2b^2 - ab^3 + b^4)$

$= a^7 + b^7 + (a + b)(a^5b - a^4b^2 + a^3b^3 - a^2b^4 + ab^5)$.   Thus, we get:

$a^7 + b^7 = (a + b)(a^6 + b^6) - (a + b)(a^5b - a^4b^2 + a^3b^3 - a^2b^4 + ab^5)$

$= (a + b)\{(a^6 + b^6) - (a^5b - a^4b^2 + a^3b^3 - a^2b^4 + ab^5)\}$

$= (a + b)(a^6 + b^6 - a^5b + a^4b^2 - a^3b^3 + a^2b^4 - ab^5)$.

Therefore, we get: $a^7 + b^7 = (a + b)(a^4 - a^5b + a^4b^2 - a^3b^3 + a^2b^4 - ab^5 + b^6)$.

Also, replacing b with $-b$ in $a^7 + b^7$, we get: $a^7 - b^7$, so we get:

$$a^7 - b^7 = (a - b)(a^4 + a^5b + a^4b^2 + a^3b^3 + a^2b^4 + ab^5 + b^6).$$

Thus, putting threads together, we get:

$$a^7 \pm b^7 = (a - b)(a^4 \mp a^5b + a^4b^2 \mp a^3b^3 + a^2b^4 \mp ab^5 + b^6).$$

How about the factorizations of $a^n \pm b^n$ and $a^m \pm b^m$ where n is odd and m is even? And of course, m and n both are integers positive.

Let's first, set: $n = 2k + 1$ and $m = 2k$ where k is a positive integer.

To begin with, putting in a sequence the polynomials of the form $a^{2k+1} \pm b^{2k+1}$, we get:

$$a^3 \pm b^3 = (a \pm b)(a^2 \mp ab + b^2).$$

$$a^5 \pm b^5 = (a \pm b)(a^4 \mp a^3b + a^2b^2 \mp ab^3 + b^4).$$

$$a^7 \pm b^7 = (a - b)(a^4 \mp a^5b + a^4b^2 \mp a^3b^3 + a^2b^4 \mp ab^5 + b^6).$$

Then, we can notice a pattern in the sequence of the exponents applied to a and b, and reflecting the pattern, we get:

$$a^{2k+1} \pm b^{2k+1} = (a \pm b)(a^{2k} \mp a^{2k-1}b + a^{2k-2}b^2 \mp a^{2k-3}b^3 + a^{2k-4}b^4 \ldots \mp ab^{2k-1} + b^{2k}).$$

Next, putting in a sequence the polynomials of the form $a^{2k} + b^{2k}$, we get:

$$a^2 + b^2 = (a + b)^2 - 2ab = (a + b + \sqrt{2ab})(a + b - \sqrt{2ab}).$$

$$a^4 + b^4 = (a^2 + b^2 + \sqrt{2}ab)(a^2 + b^2 - \sqrt{2}ab).$$

$$a^6 + b^6 = (a^3 + b^3 + \sqrt{2}a^{\frac{3}{2}}b^{\frac{3}{2}})(a^2 + b^2 - \sqrt{2}a^{\frac{3}{2}}b^{\frac{3}{2}}).$$

Then, we can notice a pattern in the sequence of the exponents applied to *a* and *b*, and reflecting the pattern, we get:

$$a^{2k} + b^{2k} = (a^k + b^k + \sqrt{2}a^{\frac{k}{2}}b^{\frac{k}{2}})(a^2 + b^2 - \sqrt{2}a^{\frac{k}{2}}b^{\frac{k}{2}}).$$

Next, putting in a sequence the polynomials of the form $a^{2k} - b^{2k}$, we get:

$$a^2 - b^2 = (a + b)(a - b).$$

$$a^4 - b^4 = (a^2 + b^2)(a + b)(a - b).$$

$$a^6 - b^6 = (a^3 + b^3)(a - b)(a^2 + ab + b^2).$$

$$a^8 - b^8 = (a^4 + b^4)(a^4 - b^4) = (a^4 + b^4)(a^2 + b^2)(a^2 - b^2) = (a^4 + b^4)(a^2 + b^2)(a + b)(a - b).$$

$$a^{10} - b^{10} = (a^5 + b^5)(a^5 - b^5), \text{ and we have: } a^5 - b^5 = (a - b)(a^4 + a^3b + a^2b^2 + ab^3 + b^4).$$

So we get: $a^{10} - b^{10} = (a^5 + b^5)(a - b)(a^4 + a^3b + a^2b^2 + ab^3 + b^4).$

Next, we get:

$$a^{12} - b^{12} = (a^6 + b^6)(a^3 + b^3)(a^3 - b^3) = (a^6 + b^6)(a^3 + b^3)(a - b)(a^2 + ab + b^2), \text{ and}$$

$$a^{14} - b^{14} = (a^7 + b^7)(a^7 - b^7) = (a^7 + b^7)(a^7 - b^7)$$
$$= (a^7 + b^7)(a - b)(a^4 + a^5b + a^4b^2 + a^3b^3 + a^2b^4 + ab^5 + b^6).$$

Next, putting in a sequence the polynomials of the form $a^n - b^n$, where $n = 2^k$, we get a sequence as follows:

$a^2 - b^2 = (a + b)(a - b).$

$a^4 - b^4 = (a^2 + b^2)(a + b)(a - b).$

$a^8 - b^8 = (a^4 + b^4)(a^2 + b^2)(a + b)(a - b).$

$a^{16} - b^{16} = (a^8 + b^8)(a^4 + b^4)(a^2 + b^2)(a + b)\ (a - b).$

Next, putting in a sequence the polynomials of the form $a^n - b^n$, where $n = 6k$, we get a sequence as follows:

$a^6 - b^6 = (a^3 + b^3)(a - b)(a^2 + ab + b^2).$

$a^{12} - b^{12} = (a^6 + b^6)(a^3 + b^3)(a - b)(a^2 + ab + b^2).$

$a^{18} - b^{18} = (a^9 + b^9)(a^6 + b^6)(a^3 + b^3)(a - b)(a^2 + ab + b^2).$

Otherwise, we get a sequence as follows:

$a^{10} - b^{10} = (a^5 + b^5)(a - b)(a^4 + a^3b + a^2b^2 + ab^3 + b^4).$

$a^{14} - b^{14} = (a^7 + b^7)(a - b)(a^6 + a^5b + a^4b^2 + a^3b^3 + a^2b^4 + ab^5 + b^6).$

$a^{20} - b^{20} = (a^{10} + b^{10})(a^5 + b^5)(a - b)(a^4 + a^3b + a^2b^2 + ab^3 + b^4).$

$a^{22} - b^{22} = (a^{11} + b^{11})(a^{11} - b^{11}).$

And we have:

$a^{2k+1} \pm b^{2k+1} = (a \pm b)(a^{2k} \mp a^{2k-1}b + a^{2k-2}b^2 \mp a^{2k-3}b^3 + a^{2k-4}b^4 \dots \mp ab^{2k-1} + b^{2k}).$

So if $k = 5$, we get:

$$a^{11} - b^{11} = (a - b)(a^{10} + a^9b + a^8b^2 + a^7b^3 + a^6b^4 + \ldots + ab^9 + b^{10}).$$

Thus, we get:

$$a^{22} - b^{22} = (a^{11} + b^{11})(a - b)(a^{10} + a^9b + a^8b^2 + a^7b^3 + a^6b^4 + \ldots + ab^9 + b^{10}).$$

Examples I

Factorize each of the following polynomials.

0. $a^3 + b^3 + c^3 - 3abc$

1. $a^4 + a^2b^2 + b^4$

Suggestions or Solutions
To the **Problem** in the Example **0**

We have: $a^3 + b^3 + c^3 - 3abc$.

Factorizing it, we can get it the way below:

To begin with, we can get:

$$(a + b + c)(a^2 + b^2 + c^2) = a^3 + b^3 + c^3 + a^2b + ab^2 + b^2c + bc^2 + c^2a + ca^2$$
$$= a^3 + b^3 + c^3 + ab(a + b) + bc(b + c) + ca(c + a)$$

And we can have:

$$ab(a + b + c) = ab^2 + a^2b + abc, \qquad bc(a + b + c) = b^2c + bc^2 + abc,$$
$$\text{and } ca(a + b + c) = c^2a + ca^2 + abc.$$

So taking the sum of the three equalities above, we get:

$$ab(a + b + c) + bc(a + b + c) + ca(a + b + c)$$
$$= ab^2 + a^2b + abc + b^2c + bc^2 + abc + c^2a + ca^2 + abc$$
$$= ab^2 + a^2b + b^2c + bc^2 + c^2a + ca^2 + 3abc.$$

And we know: $ab(a + b + c) + bc(a + b + c) + ca(a + b + c) = (a + b + c)(ab + bc + ca)$.

So we get: $(a + b + c)(ab + bc + ca) = ab^2 + a^2b + b^2c + bc^2 + c^2a + ca^2 + 3abc$.

Now, we have:

$$(a + b + c)(a^2 + b^2 + c^2) = a^3 + b^3 + c^3 + ab^2 + a^2b + b^2c + bc^2 + c^2a + ca^2, \text{ and}$$
$$(a + b + c)(ab + bc + ca) = ab^2 + a^2b + b^2c + bc^2 + c^2a + ca^2 + 3abc.$$

That is, assuming $D = ab^2 + a^2b + b^2c + bc^2 + c^2a + ca^2$, we get:

$$(a + b + c)(a^2 + b^2 + c^2) = a^3 + b^3 + c^3 + D, \text{ and } (a + b + c)(ab + bc + ca) = D + 3abc.$$

So we get:

$$(a + b + c)\{(a^2 + b^2 + c^2) - (ab + bc + ca)\}$$
$$= (a + b + c)(a^2 + b^2 + c^2) - (a + b + c)(ab + bc + ca)$$
$$= a^3 + b^3 + c^3 + D - (D + 3abc) = a^3 + b^3 + c^3 - 3abc.$$

Thus, we get: $a^3 + b^3 + c^3 - 3abc = (a + b + c)(a^2 + b^2 + c^2 - ab - bc - ca)$.

If not quite sure of the idea behind the processes above, follow the steps below:

Factorizing a polynomial, we are expecting that the polynomial has factors.
Once having found all the factors, we put them in a form of a product.
Multiplying out those factors or expanding the form, we get the polynomial back.

And such a factor can be a polynomial, and also, can be a monomial, constant, or integer.
So multiplying two polynomials, what do we get?

We get a polynomial that can be factorized.

Let's now set: $P = a^3 + b^3 + c^3 - 3abc$.
The polynomial P has four terms, which however, have nothing in common.
In other words, there is no divisor common to all the terms.

So the polynomial P looks quite far from being factorable, doesn't it?

Even having no divisor common to all the terms though, it can still be factorized.

We know multiplying two polynomials, we get a polynomial that is factorable.

Multiplying for instance, $x + 1$ by $x + 2$, we get: $x^2 + 3x + 2$, which is thus, factorable.

So if P is factorable, there must be two polynomials that can get multiplied together to produce all the four terms, a^3, b^3, c^3, and $-3abc$. And thus, we may want to think of such two polynomials.

When choosing or coming up with such polynomials though, we want to be reasonable, too, of course. That is, the product of the two should produce a^3, b^3, c^3, and $-3abc$.

For instance, we can begin with two polynomials $a + b + c$ and $a^2 + b^2 + c^2$.

Then, expanding the product of the two, we get:

$$(a + b + c)(a^2 + b^2 + c^2) = a^3 + ab^2 + ac^2 + a^2b + b^3 + bc^2 + a^2c + b^2c + c^3.$$

So we now have: $a^3 + b^3 + c^3$, but don't have: $-3abc$, and we don't want the expression as follows: $ab^2 + ac^2 + a^2b + bc^2 + a^2c + b^2c$.   What do we do then?

We can try doing something to the product so that the terms unwanted get canceled, and the term needed gets created during the expansion.   How?

We should be able to get: $-(ab^2 + ac^2 + a^2b + bc^2 + a^2c + b^2c)$ and $-3abc$, both of which need to be produced during the expansion.   How?

To begin with, rearranging the first of the two expressions above, we get:

$$-(ab^2 + ac^2 + a^2b + bc^2 + a^2c + b^2c) = -(a^2b + ab^2 + b^2c + bc^2 + c^2a + ca^2).$$

Next, looking at $a^2b + ab^2$ a bit more closely, we can notice that there is a common divisor, which is: ab.

That is to say that $a^2b + ab^2$ can be factorized to $ab(a + b)$.

And the same is true for $b^2c + bc^2$ and $c^2a + ca^2$, too.

So we get: $ab^2 + a^2b = ab(a + b)$, $b^2c + bc^2 = bc(b + c)$, and $c^2a + ca^2 = ca(c + a)$.

And let's next, move on to **-3abc**.

Looking at first, the three expressions above, we can notice that adding each of *a*, *b*, and *c* to each of the three expressions, we can get three of (*abc*)s in such a way as follows:

$$ab^2 + a^2b = ab(a + b) \Rightarrow ab(a + b + c) = ab^2 + a^2b + abc,$$

$$b^2c + bc^2 = bc(b + c) \Rightarrow bc(a + b + c) = b^2c + bc^2 + abc,$$

$$\text{and } c^2a + ca^2 = ca(c + a) \Rightarrow ca(a + b + c) = c^2a + ca^2 + abc.$$

So taking the sum of the three equalities above, we get:

$$ab(a + b + c) + bc(a + b + c) + ca(a + b + c)$$

$$= ab^2 + a^2b + abc + b^2c + bc^2 + abc + c^2a + ca^2 + abc$$

$$= ab^2 + a^2b + b^2c + bc^2 + c^2a + ca^2 + 3abc.$$

And factorizing **ab(a + b + c) + bc(a + b + c) + ca(a + b + c)**, which is the top of the equalities above, we get: **(a + b + c)(ab + bc + ca)**.

So we get: **(a + b + c)(ab + bc + ca) = ab² + a²b + b²c + bc² + c²a + ca² + 3abc**.

Now, we can see what we need.   What then, is it?

It is: **-(a + b + c)(ab + bc + ca)**.

And since **(a + b + c)(ab + bc + ca) = ab² + a²b + b²c + bc² + c²a + ca² + 3abc**, we get:

$$-(a + b + c)(ab + bc + ca)$$
$$= -(ab^2 + a^2b + b^2c + bc^2 + c^2a + ca^2 + 3abc)$$
$$= -(ab^2 + a^2b + b^2c + bc^2 + c^2a + ca^2) - 3abc.$$

So we can add **-(ab + bc + ca)** to the polynomial **a² + b² + c²**.   What then?

We know: $(a + b + c)(a^2 + b^2 + c^2) = a^3 + b^3 + c^3 + a^2b + ab^2 + b^2c + bc^2 + c^2a + ca^2$.

So on the left hand side of the equality above, adding $-(ab + bc + ca)$ to $(a^2 + b^2 + c^2)$, we get: $(a + b + c)\{(a^2 + b^2 + c^2) - (ab + bc + ca)\}$, which produces:

not only $(a + b + c)(a^2 + b^2 + c^2)$ but $-(a + b + c)(ab + bc + ca)$, too.

That is to say that we get: $-(ab^2 + a^2b + b^2c + bc^2 + c^2a + ca^2) - 3abc$ as well as

$a^3 + b^3 + c^3 + a^2b + ab^2 + b^2c + bc^2 + c^2a + ca^2$.

And thus, putting threads together, we have:

$(a + b + c)(a^2 + b^2 + c^2) = a^3 + b^3 + c^3 + ab^2 + a^2b + b^2c + bc^2 + c^2a + ca^2$, and

$(a + b + c)(ab + bc + ca) = ab^2 + a^2b + b^2c + bc^2 + c^2a + ca^2 + 3abc$.

That is, assuming $D = ab^2 + a^2b + b^2c + bc^2 + c^2a + ca^2$, we get:

$(a + b + c)(a^2 + b^2 + c^2) = a^3 + b^3 + c^3 + D$, and $(a + b + c)(ab + bc + ca) = D + 3abc$.

So we get:

$(a + b + c)\{(a^2 + b^2 + c^2) - (ab + bc + ca)\}$

$= (a + b + c)(a^2 + b^2 + c^2) - (a + b + c)(ab + bc + ca)$

$= a^3 + b^3 + c^3 + D - (D + 3abc) = a^3 + b^3 + c^3 - 3abc$.

And we know:

$(a + b + c)\{(a^2 + b^2 + c^2) - (ab + bc + ca)\} = (a + b + c)(a^2 + b^2 + c^2 - ab - bc - ca)$.

Thus, we get: $a^3 + b^3 + c^3 - 3abc = (a + b + c)(a^2 + b^2 + c^2 - ab - bc - ca)$.

In short:

To begin with, we can get:

$$(a + b + c)(a^2 + b^2 + c^2) = a^3 + b^3 + c^3 + a^2b + ab^2 + b^2c + bc^2 + c^2a + ca^2$$
$$= a^3 + b^3 + c^3 + ab(a + b) + bc(b + c) + ca(c + a)$$

And we can have:

$$ab(a + b + c) = ab^2 + a^2b + abc, \qquad bc(a + b + c) = b^2c + bc^2 + abc,$$
$$\text{and } ca(a + b + c) = c^2a + ca^2 + abc.$$

So taking the sum of the three equalities above, we get:

$$ab(a + b + c) + bc(a + b + c) + ca(a + b + c)$$
$$= ab^2 + a^2b + abc + b^2c + bc^2 + abc + c^2a + ca^2 + abc$$
$$= ab^2 + a^2b + b^2c + bc^2 + c^2a + ca^2 + 3abc.$$

And we know: $ab(a + b + c) + bc(a + b + c) + ca(a + b + c) = (a + b + c)(ab + bc + ca)$.

So we get: $(a + b + c)(ab + bc + ca) = ab^2 + a^2b + b^2c + bc^2 + c^2a + ca^2 + 3abc$.

Now, we have:

$$(a + b + c)(a^2 + b^2 + c^2) = a^3 + b^3 + c^3 + ab^2 + a^2b + b^2c + bc^2 + c^2a + ca^2, \text{ and}$$
$$(a + b + c)(ab + bc + ca) = ab^2 + a^2b + b^2c + bc^2 + c^2a + ca^2 + 3abc.$$

That is, assuming $D = ab^2 + a^2b + b^2c + bc^2 + c^2a + ca^2$, we get:

$$(a + b + c)(a^2 + b^2 + c^2) = a^3 + b^3 + c^3 + D, \text{ and } (a + b + c)(ab + bc + ca) = D + 3abc.$$

So we get:
$$(a + b + c)\{(a^2 + b^2 + c^2) - (ab + bc + ca)\}$$
$$= (a + b + c)(a^2 + b^2 + c^2) - (a + b + c)(ab + bc + ca)$$
$$= a^3 + b^3 + c^3 + D - (D + 3abc) = a^3 + b^3 + c^3 - 3abc.$$

Thus, we get: $a^3 + b^3 + c^3 - 3abc = (a + b + c)(a^2 + b^2 + c^2 - ab - bc - ca)$.

Suggestions or Solutions
To the **Problem** in the Example **1**

We have: $a^4 + a^2b^2 + b^4$.

Factorizing it, we can get it the way below:

$(a^2 + b^2)^2 - a^2b^2 = a^4 + 2a^2b^2 + b^4 - a^2b^2 = a^4 + a^2b^2 + b^4$, and also,

$(a^2 + b^2)^2 - a^2b^2 = \{(a^2 + b^2) + ab\}\{(a^2 + b^2) - ab\} = (a^2 + b^2 + ab)(a^2 + b^2 - ab)$.

Therefore, $a^4 + a^2b^2 + b^4 = (a^2 + b^2 + ab)(a^2 + b^2 - ab)$.

If not quite sure of the idea behind the processes above, follow the steps below:

First, we know $a^2 + 2ab + b^2$ gets factorized to $(a + b)^2$.

So we may want to begin with the expansion of $(a^2 + b^2)^2$:

Then, we get:

$(a^2 + b^2)^2 = a^4 + 2a^2b^2 + b^4$, which is $a^4 + a^2b^2 + a^2b^2 + b^4$, in which however, we have too many of (a^2b^2)s, since we need only one of those.

So next, we want to compensate a^2b^2 so that we can have: $a^4 + a^2b^2 + b^4$ only.   How?

We have: $(a^2 + b^2)^2 = a^4 + 2a^2b^2 + b^4 = a^4 + a^2b^2 + a^2b^2 + b^4$, so we want to compensate a^2b^2.   What then, can we do?

We can do this: $(a^4 + a^2b^2 + a^2b^2 + b^4) + (-a^2b^2)$.

Then, we get: $a^4 + a^2b^2 + a^2b^2 + b^4 - a^2b^2$, and thus, we can get: $a^4 + a^2b^2 + b^4$.

Now, we have: $(a^2 + b^2)^2 = a^4 + a^2b^2 + a^2b^2 + b^4$.

(Doing one thing to the right hand side of an equality, we want to do the same to the left hand side, too, if we want to maintain the equality.)

So we may want to try $(a^2 + b^2)^2 - a^2b^2$.
We are not going to just do the subtraction only, but take advantage of a factorization identity as follows, too: $x^2 - y^2 = (x + y)(x - y)$.

Then, we can get:

$(a^2 + b^2)^2 - a^2b^2 = \{(a^2 + b^2) + ab\}\{(a^2 + b^2) - ab\} = (a^2 + b^2 + ab)(a^2 + b^2 - ab)$ as well as $a^4 + 2a^2b^2 + b^4 - a^2b^2 = a^4 + a^2b^2 + a^2b^2 + b^4 - a^2b^2 = a^4 + a^2b^2 + b^4$.

That is to say that we get:

$(a^2 + b^2)^2 - a^2b^2 = \{(a^2 + b^2) + ab\}\{(a^2 + b^2) - ab\} = \underline{(a^2 + b^2 + ab)(a^2 + b^2 - ab)}$, and

$(a^2 + b^2)^2 - a^2b^2 = a^4 + 2a^2b^2 + b^4 - a^2b^2 = a^4 + a^2b^2 + a^2b^2 + b^4 - a^2b^2 = \underline{a^4 + a^2b^2 + b^4}$.

Therefore, we get: $a^4 + a^2b^2 + b^4 = (a^2 + b^2 + ab)(a^2 + b^2 - ab)$.

In short:

$(a^2 + b^2)^2 - a^2b^2 = a^4 + 2a^2b^2 + b^4 - a^2b^2 = a^4 + a^2b^2 + b^4$, and also,

$(a^2 + b^2)^2 - a^2b^2 = \{(a^2 + b^2) + ab\}\{(a^2 + b^2) - ab\} = (a^2 + b^2 + ab)(a^2 + b^2 - ab)$.

Therefore, $a^4 + a^2b^2 + b^4 = (a^2 + b^2 + ab)(a^2 + b^2 - ab)$.

84

Examples J

0.   Factorize each of the polynomials below:

0.0. $a^3 + 3a^2b + 3ab^2 + b^3$

0.1. $a^3 - 3a^2b + 3ab^2 - b^3$

1.   Show that $a^3 + b^3 + c^3 - 3abc = \frac{1}{2}(a + b + c)\{(a - b)^2 + (b - c)^2 + (c - a)^2\}$.

2.   Assuming that $a + b = 19$, and that $ab = 29$, find the value of $a^3 + b^3$.

Suggestions or Solutions
To the **Problem 0** in the Example **0**

We have: $a^3 + 3a^2b + 3ab^2 + b^3$.

Factorizing it, we can get it the way below:

$a^3 + 3a^2b + 3ab^2 + b^3 = a^3 + a^2b + 2a^2b + 2ab^2 + ab^2 + b^3$
$= a^2(a + b) + 2ab(a + b) + b^2(a + b) = (a + b)(a^2 + 2ab + b^2) = (a + b)(a + b)^2 = (a + b)^3$.

$a^3 + 3a^2b + 3ab^2 + b^3 = 3a^2b + 3ab^2 + a^3 + b^3 = 3ab(a + b) + a^3 + b^3$
$= 3ab(a + b) + (a + b)(a^2 - ab + b^2) = (a + b)(3ab + a^2 - ab + b^2)$
$= (a + b)(a^2 + 2ab + b^2) = (a + b)(a + b)^2 = (a + b)^3$.

If not quite sure of the idea behind the processes above, follow the steps below:

Let's set first, $P = a^3 + 3a^2b + 3ab^2 + b^3$.

Basically, factorizing a polynomial, we begin with finding a common divisor.
Such a divisor can be common to all or some of the polynomial.
So if not able to find a divisor common to all the terms, we may want to try finding a divisor common to some of the terms.

Now, we can put P this way, too: $a^3 + a^2b + 2a^2b + 2ab^2 + ab^2 + b^3$.

Then, we can see that a^2 is common to a^3 and a^2b, $2ab$ is common to $2a^2b$ and $2ab^2$, and b^2 is common to ab^2 and b^3.

So we get:

$P = a^3 + a^2b + 2a^2b + 2ab^2 + ab^2 + b^3 = a^2(a + b) + 2ab(a + b) + b^2(a + b)$, where $a + b$ is common.

Thus, we get: $P = (a + b)(a^2 + 2ab + b^2) = (a + b)(a + b)^2 = (a + b)^3$.

We can get the same the way below, too:

We know that $a^3 + b^3$ gets factorized to $(a + b)(a^2 - ab + b^2)$, and that $3a^2b + 3ab^2$ can get factorized to $3ab(a + b)$.

So $a + b$ can be said to be common to $a^3 + b^3$ and $3a^2b + 3ab^2$.   Thus, we can get:

$P = a^3 + 3a^2b + 3ab^2 + b^3 = 3a^2b + 3ab^2 + a^3 + b^3 = 3ab(a + b) + a^3 + b^3$

$= 3ab(a + b) + (a + b)(a^2 - ab + b^2) = (a + b)(3ab + a^2 - ab + b^2)$

$= (a + b)(a^2 + 2ab + b^2) = (a + b)(a + b)^2 = (a + b)^3$.

In short:

By one method:

$a^3 + 3a^2b + 3ab^2 + b^3 = a^3 + a^2b + 2a^2b + 2ab^2 + ab^2 + b^3$

$= a^2(a + b) + 2ab(a + b) + b^2(a + b) = (a + b)(a^2 + 2ab + b^2) = (a + b)(a + b)^2 = (a + b)^3$.

And by the other:

$a^3 + 3a^2b + 3ab^2 + b^3 = 3a^2b + 3ab^2 + a^3 + b^3 = 3ab(a + b) + a^3 + b^3$

$= 3ab(a + b) + (a + b)(a^2 - ab + b^2) = (a + b)(3ab + a^2 - ab + b^2)$

$= (a + b)(a^2 + 2ab + b^2) = (a + b)(a + b)^2 = (a + b)^3$.

Suggestions or Solutions
To the **Problem 1** in the Example **0**

We have: $a^3 - 3a^2b + 3ab^2 - b^3$.

Factorizing it, we can get it the way below:

$a^3 - 3a^2b + 3ab^2 - b^3 = a^3 - a^2b - 2a^2b + 2ab^2 + ab^2 - b^3$

$= a^2(a - b) - 2ab(a - b) + b^2(a - b) = (a - b)(a^2 - 2ab + b^2) = (a - b)(a - b)^2 = (a - b)^3$.

$a^3 - 3a^2b + 3ab^2 - b^3 = -3a^2b + 3ab^2 + a^3 - b^3 = -3ab(a - b) + a^3 - b^3$

$= -3ab(a - b) + (a - b)(a^2 + ab + b^2) = (a - b)(-3ab + a^2 + ab + b^2)$

$= (a - b)(a^2 - 2ab + b^2) = (a - b)(a - b)^2 = (a - b)^3$.

If not quite sure of the idea behind the processes above, follow the steps below:

Let's set first, $P = a^3 - 3a^2b + 3ab^2 - b^3$.

Basically, factorizing a polynomial, we find a common divisor.
Such a divisor can be common to all or some of the polynomial.

So if not able to find a divisor common to all the terms, we may want to try finding a divisor common to some of the terms.

Now, we can put P this way, too: $P = a^3 - a^2b - 2a^2b + 2ab^2 + ab^2 - b^3$.

Then, we can see that a^2 is common to a^3 and $-a^2b$, $2ab$ is common to $-2a^2b$ and $2ab^2$, and b^2 is common to ab^2 and $-b^3$. So we get:

$P = a^3 - a^2b - 2a^2b + 2ab^2 + ab^2 - b^3 = a^2(a - b) - 2ab(a - b) + b^2(a - b)$, where $a - b$ is common. Thus, we get: $P = (a - b)(a^2 - 2ab + b^2) = (a - b)(a - b)^2 = (a - b)^3$.

• And we can get the same the way below, too:

We know that $a^3 - b^3$ gets factorized to $(a - b)(a^2 + ab + b^2)$, and that $3a^2b - 3ab^2$ can get factorized to $3ab(a - b)$.

So $a - b$ can be said to be common to $a^3 - b^3$ and $3a^2b - 3ab^2$.

Thus, we can get:

$$P = a^3 - 3a^2b + 3ab^2 - b^3 = \text{-}3a^2b + 3ab^2 + a^3 - b^3 = \text{-}3ab(a - b) + a^3 - b^3$$

$$= \text{-}3ab(a - b) + (a - b)(a^2 + ab + b^2) = (a - b)(\text{-}3ab + a^2 + ab + b^2)$$

$$= (a - b)(a^2 - 2ab + b^2) = (a - b)(a - b)^2 = (a - b)^3.$$

And of course, replacing b with $\text{-}b$, too, we get: $a^3 - 3a^2b + 3ab^2 - b^3 = (a - b)^3$.

In short:

By one method:

$$a^3 - 3a^2b + 3ab^2 - b^3 = a^3 - a^2b - 2a^2b + 2ab^2 + ab^2 - b^3$$

$$= a^2(a - b) - 2ab(a - b) + b^2(a - b) = (a - b)(a^2 - 2ab + b^2) = (a - b)(a - b)^2 = (a - b)^3.$$

And by the other:

$$a^3 - 3a^2b + 3ab^2 - b^3 = \text{-}3a^2b + 3ab^2 + a^3 - b^3 = \text{-}3ab(a - b) + a^3 - b^3$$

$$= \text{-}3ab(a - b) + (a - b)(a^2 + ab + b^2) = (a - b)(\text{-}3ab + a^2 + ab + b^2)$$

$$= (a - b)(a^2 - 2ab + b^2) = (a - b)(a - b)^2 = (a - b)^3.$$

Suggestions or Solutions
To the **Problem** in the Example 1

Show that $a^3 + b^3 + c^3 - 3abc = \frac{1}{2}(a + b + c)\{(a - b)^2 + (b - c)^2 + (c - a)^2\}$.

Quite frequently, we need to show proofs doing problems in math.
In particular, showing that an equality holds, that is, proving an equality, we need to show two facts. We want to show not only that one side of the equality implies the other, but that the other implies the one side, too.

So for instance, showing that $A = B$, we have to show B implies A as well as A implies B. In short, we must show $A \Rightarrow B$ and $B \Rightarrow A$ both. So both directions have to be OK.

Why both, though?

Sometimes, the other way does not work. That is to say that in some cases, the other way is not always true. For instance, we can say that bread is food, but cannot say that food is bread because some food is not bread. For instance, an apple is food, but is not bread. So in math, it is not true that **bread = food**. That is, **bread ≠ food**, in math. What if it were true that bread = food, though?

Then, all food would have to be bread.

So proving an equality as a factorization identity, we want to show both ways: one way and the other.   In this case though, it is not hard to show one way, which is:

$\frac{1}{2}(a + b + c)\{(a - b)^2 + (b - c)^2 + (c - a)^2\}$ implies $a^3 + b^3 + c^3 - 3abc$.

Just expanding (simplifying) the right hand side of the equality, we get the left hand side. Let's expand it, and see if it is the case.   To begin with, we can get:

$$(a - b)^2 + (b - c)^2 + (c - a)^2\} = a^2 - 2ab + b^2 + b^2 - 2bc + c^2 + c^2 - 2ca + a^2$$

$$= 2(a^2 + b^2 + c^2 - ab - bc - ca).$$

So next, we get:

$$\tfrac{1}{2}(a + b + c)\{(a - b)^2 + (b - c)^2 + (c - a)^2\} = (a + b + c)(a^2 + b^2 + c^2 - ab - bc - ca)$$

$$= a^3 + ab^2 + ac^2 - a^2b - abc - a^2c + ba^2 + b^3 + bc^2 - ab^2 - b^2c - abc + ca^2 + cb^2 + c^3 - abc - bc^2 - c^2a = a^3 + b^3 + c^3 - 3abc.$$

So such an expansion and simplification show us:

$$\tfrac{1}{2}(a + b + c)\{(a - b)^2 + (b - c)^2 + (c - a)^2\} \Rightarrow a^3 + b^3 + c^3 - 3abc.$$

However, it may not be quite straightforward to show that the other way is true. That is to say that it is not quite simple to show that:

$$a^3 + b^3 + c^3 - 3abc \Rightarrow \tfrac{1}{2}(a + b + c)\{(a - b)^2 + (b - c)^2 + (c - a)^2\}.$$

Memorizing the identity $a^3 + b^3 + c^3 - 3abc = (a + b + c)(a^2 + b^2 + c^2 - ab - bc - ca)$ can be of significant help, but we still need to show that:

$$(a + b + c)(a^2 + b^2 + c^2 - ab - bc - ca) \Rightarrow \tfrac{1}{2}(a + b + c)\{(a - b)^2 + (b - c)^2 + (c - a)^2\}.$$

To make it a bit simpler, we can just show that:

$$a^2 + b^2 + c^2 - ab - bc - ca \Rightarrow \tfrac{1}{2}\{(a - b)^2 + (b - c)^2 + (c - a)^2\}, \text{ since } (a + b + c) \text{ is}$$

common to both of the sides.

Now, compensating, we can get:

$$a^2 + b^2 + c^2 - ab - bc - ca = \tfrac{1}{2} \cdot 2(a^2 + b^2 + c^2 - ab - bc - ca)$$

$$= \tfrac{1}{2}(2a^2 + 2b^2 + 2c^2 - 2ab - 2bc - 2ca) = \tfrac{1}{2}(a^2 - 2ab + b^2 + b^2 - 2bc + c^2 + c^2 - 2ca + a^2)$$

$$= \tfrac{1}{2}\{(a - b)^2 + (b - c)^2 + (c - a)^2\}.$$

So we get: $a^2 + b^2 + c^2 - ab - bc - ca \Rightarrow \tfrac{1}{2}\{(a - b)^2 + (b - c)^2 + (c - a)^2\}$, and in turn, we get: $a^3 + b^3 + c^3 - 3abc \Rightarrow \tfrac{1}{2}(a + b + c)\{(a - b)^2 + (b - c)^2 + (c - a)^2\}.$

Therefore, we can now say that:

$$a^3 + b^3 + c^3 - 3abc = \tfrac{1}{2}(a + b + c)\{(a - b)^2 + (b - c)^2 + (c - a)^2\}.$$

In short:

To begin with, showing $\tfrac{1}{2}(a + b + c)\{(a - b)^2 + (b - c)^2 + (c - a)^2\} \Rightarrow a^3 + b^3 + c^3 - 3abc$,

we get first: $(a - b)^2 + (b - c)^2 + (c - a)^2\} = a^2 - 2ab + b^2 + b^2 - 2bc + c^2 + c^2 - 2ca + a^2$

$= 2(a^2 + b^2 + c^2 - ab - bc - ca)$.

So $\tfrac{1}{2}(a + b + c)\{(a - b)^2 + (b - c)^2 + (c - a)^2\} = (a + b + c)(a^2 + b^2 + c^2 - ab - bc - ca)$

$= a^3 + ab^2 + ac^2 - a^2b - abc - a^2c + ba^2 + b^3 + bc^2 - ab^2 - b^2c - abc + ca^2 + cb^2 + c^3 -$

$abc - bc^2 - c^2a = a^3 + b^3 + c^3 - 3abc$.

So we get: $\tfrac{1}{2}(a + b + c)\{(a - b)^2 + (b - c)^2 + (c - a)^2\} \Rightarrow a^3 + b^3 + c^3 - 3abc$.

Next, we want to show: $a^3 + b^3 + c^3 - 3abc \Rightarrow \tfrac{1}{2}(a + b + c)\{(a - b)^2 + (b - c)^2 + (c - a)^2\}$.

To begin with, we have an identity as follows:
$a^3 + b^3 + c^3 - 3abc = (a + b + c)(a^2 + b^2 + c^2 - ab - bc - ca)$.

So we want to show:
$(a + b + c)(a^2 + b^2 + c^2 - ab - bc - ca) \Rightarrow \tfrac{1}{2}(a + b + c)\{(a - b)^2 + (b - c)^2 + (c - a)^2\}$.

Thus, we want to show: $(a^2 + b^2 + c^2 - ab - bc - ca) \Rightarrow \tfrac{1}{2}\{(a - b)^2 + (b - c)^2 + (c - a)^2\}$.

Now, compensating, we get: $a^2 + b^2 + c^2 - ab - bc - ca = \tfrac{1}{2} \cdot 2(a^2 + b^2 + c^2 - ab - bc - ca)$

$= \tfrac{1}{2}(2a^2 + 2b^2 + 2c^2 - 2ab - 2bc - 2ca) = \tfrac{1}{2}(a^2 - 2ab + b^2 + b^2 - 2bc + c^2 + c^2 - 2ca + a^2)$

$= \tfrac{1}{2}\{(a - b)^2 + (b - c)^2 + (c - a)^2\}$.

So we get: $a^3 + b^3 + c^3 - 3abc \Rightarrow \tfrac{1}{2}(a + b + c)\{(a - b)^2 + (b - c)^2 + (c - a)^2\}$.

Therefore, $a^3 + b^3 + c^3 - 3abc = \tfrac{1}{2}(a + b + c)\{(a - b)^2 + (b - c)^2 + (c - a)^2\}$.

Suggestions or Solutions
To the **Problem** in the Example **2**

Assuming that $a + b = 19$, and that $ab = 29$, find the value of $a^3 + b^3$.

$$(a + b)^3 = a^3 + 3ab(a + b) + b^3 \Rightarrow a^3 + b^3 = (a + b)^3 - 3ab(a + b) = 19^3 - 3 \cdot 29 \cdot 19$$

$$= 6859 - 1653 = 5206.$$

If not quite sure of the idea behind the processes above, follow the steps below:

Solving the system of equations where $a + b = 19$ and $ab = 29$, we can get the values of a and b. And then, putting the values into $a^3 + b^3$, we can get the value of $a^3 + b^3$.

Solving the system is though, nothing but solving the quadratic equation below:

$x^2 - 19x + 29 = 0$.   How come?

That's because: $x^2 - (a + b)x + ab = (x - a)(x - b) = 0 \Rightarrow x = a$ or b.

And we have: $a + b = 19$, and $ab = 29$.

So if we can factorize $x^2 - 19x + 29$, we can get the solution.

And thus, the values of a and b are the roots of the equation $x^2 - 19x + 29 = 0$.

We have however, $19^2 - 4 \cdot 1 \cdot 29 = 245$, which is called the discriminant, and is not an integer squared. So the roots are not simple numbers as integers or rational numbers but irrational numbers.

That is, $x^2 - 19x + 29$ is not factorized to $(x - m)(x - n)$ where m and n are integers.

So the roots are numbers not simple enough. Thus, if we put them directly into $a^3 + b^3$, it can hardly be the case where we can get the value of $a^3 + b^3$ fast enough.

Using in fact, the quadratic formula, we get:

$x = \frac{19 \pm \sqrt{19^2 - 4 \cdot 1 \cdot 29}}{2} = \frac{19 \pm \sqrt{245}}{2} = \frac{19 \pm \sqrt{5 \cdot 49}}{2} = \frac{19 \pm 7\sqrt{5}}{2}$, which are the values of a and b.

And taking the cubes of those two, we get numbers quite complicated.
We can get around with such a mess though, using a factorization identity.
What factorization identity though?

We have a couple of identities as follows:

$a^3 + b^3 = (a + b)(a^2 - ab + b^2)$ and $(a + b)^3 = a^3 + 3ab(a + b) + b^3$.

We can use either of the two above, but the second one is better.
So using the second, we get:

$(a + b)^3 = a^3 + 3ab(a + b) + b^3 \Rightarrow a^3 + b^3 = (a + b)^3 - 3ab(a + b) = 19^3 - 3 \cdot 29 \cdot 19$

$= 6859 - 1653 = 5206$.

Let's next, use the first identity.

Then, we want to get the value of $a^2 + b^2$, first, which can be obtained from another identity where $(a + b)^2 = a^2 + 2ab + b^2$.

Then, we get first:

$a^2 + b^2 = (a + b)^2 - 2ab = 19^2 - 2 \cdot 29 = (20 - 1)^2 - 58 = 400 - 40 + 1 - 58$

$= 301 + 100 - 98 = 303$.

So next, we get: $a^3 + b^3 = (a + b)(a^2 - ab + b^2) = 19(303 - 29) = 19 \cdot 274 = 5206$.

Examples K

0. Do the arithmetic operations below:

0.0. $99 \cdot 33 + 44 \cdot 45$

0.1. $88888888 \cdot 22222222 + 44444444 \cdot 55555556$

0.2. $999999999 \cdot 999999999 + 1999999999$, and

0.3. $\dfrac{123454321}{123454322^2 - 123454321 \cdot 123454323}$

1. Suppose A is a positive integer that has 2100 digits, and every digit is 9. So all the digits in A are 9s as 9999… Suppose also, another integer B has 2101 digits, and in B, the highest digit is 1, and all the other digits are all 9s. Then, find the value of $A^2 + B$.

2.0. Show that 1000000001 is a composite integer.

2.1. Suppose A is a positive integer as 1000…001, and has 2323 digits. So in A, the highest and the lowest digits are 1, and all the other digits are all 0s. Then, show that A is a composite integer.

Suggestions or Solutions
To the **Problems** in the Example **0**

0.0. $99 \cdot 33 + 44 \cdot 45$

0.1. $88888888 \cdot 22222222 + 44444444 \cdot 55555556$

0.2. $999999999 \cdot 999999999 + 1999999999$, and

0.3. $\dfrac{123454321}{123454322^2 - 123454321 \cdot 123454323}$

Let's begin with $99 \cdot 33 + 44 \cdot 45$.
Taking the sum above, we can take many ways. How many though?

There can be infinitely many ways. Of such many, we can choose one, and it all depends on what's most convenient to us.

$99 \cdot 33 + 44 \cdot 45$

$= (90 + 9)33 + (40 + 4)45$

$= (90 + 9)(30 + 3) + (40 + 4)(40 + 5)$

$= 2700 + 270 + 270 + 27 + 1600 + 200 + 160 + 20$

$= 2700 + 270 \cdot 2 + 27 + 1800 + 180$

$= 2700 + 540 + 27 + 1980$

$= 2700 + 300 + 240 + 7 + 20 + 1980$

$= 3000 + 240 + 7 + 2000 = 5247$

$99 \cdot 33 + 44 \cdot 45$

$= (100 - 1)33 + (40 + 4)(40 + 5)$

$= 3300 - 33 + 1600 + 200 + 160 + 20$

$= 3000 + 1000 + 300 + 600 + 200 + 100 + 60 + 20 - 30 - 3$

$= 4000 + 1200 + 60 - 10 - 3$

$= 5200 + 50 - 3 = 5200 + 47 = 5247$

Of course, we can do calculation by heart in some steps presented above.

Next, we have: 88888888·22222222 + 44444444·55555556.
Assuming first, $a = 11111111$, we get:

88888888·22222222 + 44444444·55555556
$= 8a·2a + 4a(5a + 1)$
$= 4a·4a + 4a(5a + 1)$
$= 4a(4a + 5a + 1)$
$= 4a(9a + 1)$
$= 44444444·(99999999 + 1)$
$= 44444444·10^8$, which looks better, and is more readable than 4444444400000000.

Next, we have 999999999·999999999 + 1999999999.
To begin with, assuming that $b = 999999999$, we get:
999999999·999999999 + 1999999999
$= b^2 + 1000000000 + b$
$= b^2 + 10^9 + b$
$= b(b + 1) + 10^9$
$= b10^9 + 10^9$ because $b + 1 = 999999999 + 1 = 10^9$.
$= 10^9(b + 1)$
$= 10^9 10^9$
$= (10^9)^2$
$= 10^{18}$.

And next, we have: $\dfrac{123454321}{123454322^2 - 123454321·123454323}$.

Setting first, $c = 123454321$, we get:

$$\frac{123454321}{123454322^2 - 123454321·123454323} = \frac{a}{(a+1)^2 - a(a+2)} = \frac{a}{a^2 + 2a + 1 - a^2 - 2a} = a = 123454321.$$

Suggestions or Solutions
To the **Problem** in the Example 1

Suppose A is a positive integer that has 2100 digits, and every digit is 9. So its all digits are 9s as 9999... Suppose also, another integer B has 2101 digits, the highest digit in it is 1, and all the other digits are all 9s. Then, find the value of $A^2 + B$.

To begin with, $B = 10^{2100} + A$.

So we get: $A^2 + B = A^2 + 10^{2100} + A = A(A + 1) + 10^{2100}$.

Meanwhile, $A + 1 = 10^{2100}$.

So we get:

$A^2 + B = A \cdot 10^{2100} + 10^{2100} = 10^{2100}(A + 1) = 10^{2100} 10^{2100} = (10^{2100})^2 = 10^{4200}$.

If not quite sure of the idea behind the processes above, follow the steps below:

Let's begin with B.

The highest digit in the integer B is 1, and B has one more digit than A, where all the digits are **9**s.

So B is 1999...9, where there are two thousand and one hundred 9s.

So we can put B in terms of A in such a way as follows: $B = 10^{2100} + A$.

Thus, we get: $A^2 + B = A^2 + 10^{2100} + A = A(A + 1) + 10^{2100}$.

Meanwhile, $A + 1 = 10^{2100}$, since $A = 999...999$, which has: 2100 of 9s.
How come?

We know A has 2100 of 9s, so if we add 1 to A, all of 2100 of 9s will become 2100 of zeros, and the highest digit of the sum $(A + 1)$ is 1.

Therefore, $A + 1 = 10^{2100}$, which has 2100 zeros.

So we get: $A^2 + B = A \cdot 10^{2100} + 10^{2100} = 10^{2100}(A + 1) = 10^{2100}10^{2100} = (10^{2100})^2 = 10^{4200}$.

Note that 10^{4200} has 4200 zeros.

In short:

To begin with, $B = 10^{2100} + A$.

So we get: $A^2 + B = A^2 + 10^{2100} + A = A(A + 1) + 10^{2100}$.

Meanwhile, $A + 1 = 10^{2100}$.

So we get:

$$A^2 + B = A \cdot 10^{2100} + 10^{2100} = 10^{2100}(A + 1) = 10^{2100}10^{2100} = (10^{2100})^2 = 10^{4200}.$$

Suggestions or Solutions
To the **Problems** in the Example **2**

2.0. **Show that 1000000001 is a composite integer.**

2.1. **Suppose A is a positive integer as 1000...001, and has 2323 digits. So in A, the highest and the lowest digits are 1, and all the other digits are all 0s. Then, show that A is a composite integer.**

Saying composite numbers or prime numbers, we normally mean integers, so we mean integers composite or prime.

And just saying primes or composite numbers, they are assumed to be positive integers.

What then, about -3 and -4?

-3 is a negative prime, and -4 is a negative composite integer.

A composite integer has to be a product of two different integers that are neither 1 nor the integer itself. So composite integers are positive integers except 1 and primes.

2.0. We have: $1000000001 = 1000000000 + 1 = 10^9 + 1 = (10^3)^3 + 1^3$.

Also, we have: $a^3 + b^3 = (a + b)(a^2 - ab + b^2)$.

So we get: $(10^3)^3 + 1^3 = (10^3 + 1)((10^3)^2 - 10^3 \cdot 1 + 1^2) = 1001(10^6 - 10^3 + 1)$.

We know 1001 and $10^6 - 10^3 + 1$ both are integers.

So 1000000001 is a product of two different integers that are neither 1 nor 1000000001, and thus, is a composite integer.

2.1. **Suppose** A **is a positive integer as 1000…001, and has 2323 digits. So in** A**, the highest and the lowest digits are 1, and all the other digits are all 0s. Then, show that** A **is a composite integer.**

To begin with, A has 2,323 digits and the highest and the lowest digit are 1s, and all of the other digits are all 0s.

So A looks like 1000000……001, and more specifically:

A = **1 + 1000…000** that has 2322 zeros.

Thus, A = **1000…001**, and there are 2321 zeros between the first 1 and the last 1.

So we can set: $A = 10^{2322} + 1$.

Now, since **2322 = 3·774**, we get: $A = 10^{3·774} + 1 = (10^{774})^3 + 1$.

So assuming that $B = 10^{774}$, we get:

$$A = B^3 + 1^3 = (B + 1)(B^2 - B·1 + 1^2) = (B + 1)(B^2 - B + 1)$$

$$= (10^{774} + 1)(10^{774·2} - 10^{774} + 1) = (10^{774} + 1)(10^{1548} - 10^{774} + 1).$$

We know $10^{774} + 1$ and $10^{1548} - 10^{774} + 1$ are integers.

So A is a product of two different integers that are neither 1 nor itself, and therefore, is a composite integer.

GCD and LCM 1

To begin with, what is GCD?

It is the acronym of Greatest Common Divisor. So it is a divisor common and the largest. And since a factor is a divisor, GCD is often called GCF, too, which is the acronym of Greatest Common Factor.

Considering GCD, we usually work with integers, constants, monomials, or polynomials.

Talking about GCD of integers, for instance, since it is common, we work with a group of integers, and not just one. It is common to all the integers in a group. And also, since it is a divisor, it divides every integer in the group.   What's the next, then?

It is the greatest, so it is the largest divisor that can divide every integer in the group. In other words, it is the greatest of all divisors common to all the integers in the group. And the same is true, too, for a set of constants, a set of monomials, or a set of polynomials.

Why GCD, though?

Suppose for instance, we want to distribute to students 576 apples, 360 pears, 180 bananas, and 252 oranges. Suppose also, we want to distribute all the fruit evenly in each kind. In other words, every recipient gets the same amount for each fruit. So amounts in kinds are different, but the total amount each recipient gets is the same.

Then, we can have a number of options.

Obviously, one student can get them all. So technically, it is possible. Practically though:

• Two students can be recipients, and each student gets 288 apples, 180 pears, 90 bananas, and 126 oranges.
• Three students can receive, and each student gets 192 apples, 120 pears, 60 bananas, and 84 oranges.
• Four students can receive, and each student gets 144 apples, 90 pears, 45 bananas, and 63 oranges.
And so forth.   What then, is the number of students in each case above?

The number of students is a divisor common to the numbers in kinds of fruit.
That is, the number of recipients is a divisor common to 576, 360, 180, and 252.
What then, is the number of the options we can take?

The number of options is the number of common divisors.
What if though, we want to maximize the number of recipients?

Then, we want to find the greatest of all the divisors common to all the numbers in fruit, which are integers, of course.   The greatest is the GCD.   How can we find it, though?

Anyway, the GCD is a common divisor, so first, we want to find divisors common to all the integers. Finding divisors common or not though, we should not just do divisions. Just doing divisions, we probably see divisors getting messed up. So what should we do?

We can find divisors efficiently using factors of each integer.

So prior to common divisors, we want to find factors of all the integers. Thus, we want to begin with factorizing all the integers.   Then, we get:

$576 = 2^6 3^2$, $360 = 2^3 3^2 5$, $180 = 2^2 3^2 5$, and $252 = 2^2 3^2 7$.

What is the difference though, between a divisor and a factor?

A factor is a divisor, which is not always a factor, though. Divisors of an integer can be 1 and the integer itself, too, but usually, we do not take such numbers as factors.

So for instance, a prime as 2 or 5 is often said to have no factor, because divisors of a prime are 1 and the prime itself only, and there is no point factorizing a prime.

It is quite often the case though, 1 and the integer itself are taken as factors, too. That is to say that all divisors are often taken as factors. So just saying factors of an integer, we mean divisors of the integer.

- Now, we have: $576 = 2^6 3^2$, $360 = 2^3 3^2 5$, $180 = 2^2 3^2 5$, and $252 = 2^2 3^2 7$.

So all the integers have been fully factorized, we want to get to the GCD.

Once an integer has been fully factorized, the integer is expressed in a form of a product of primes, which are prime factors, often just called factors, too.

In short, an integer factorized is a product of its factors prime.

So for instance, 576 factorized is $2^6 3^2$.

And if a set of integers share a factor, the factor is said to be common to all the integers in the set, and thus, is called a common factor. So what do we do with a common factor?

We know a factor is a divisor.
So a common factor of a set of integers can divide all the integers in the set.

Thus, we want to find factors common to all the integers. A factor begins with a prime factor.

So we want to begin with prime factors common to all the integers, which are in this case, all the numbers in fruit.

Examining all the numbers factorized, and beginning with the smallest, which is 2 in this case, we can see all common factors can be put in a sequence as follows:

$$2^1 3^0,\ 2^0 3^1,\ 2^2 3^0,\ 2^1 3^1,\ 2^0 3^2,\ 2^2 3^1,\ 2^1 3^2,\ \text{and}\ 2^2 3^2.$$

Thus, $2^2 3^2$ is the greatest of all the factors common to all the numbers, that is, all the integers $576 = 2^6 3^2$, $360 = 2^3 3^2 5$, $180 = 2^2 3^2 5$, and $252 = 2^2 3^2 7$.

So the GCD is $2^2 3^2$, which is 36. Therefore, 36 students can be recipients at the most. How much fruit then, each of those 36 students can get?

Setting GCD = G, we can put the integers the way below, too:

Apples:	$576 = 2^6 3^2 = 2^2 3^2 2^4 = 16G.$
Pears:	$360 = 2^3 3^2 5 = 2^2 3^2 2 \cdot 5 = 10G.$
Bananas:	$180 = 2^2 3^2 5 = 5G.$
Oranges:	$252 = 2^2 3^2 7 = 7G.$

And thus, each student gets 16 apples, 10 pears, 5 bananas, and 7 oranges.

So distributing objects evenly in each kind, and maximizing the recipients, we want to find the GCD of the numbers of objects in all kinds. In other words, distributing objects equally in each kind, and minimizing each number of objects in each kind each recipient gets, we want to find the GCD of the numbers of objects in all kinds.

Now, what then, about LCD? That is, what can we say about Least Common Divisor?

Technically, 1 is the LCD because 1 is the smallest (positive) of all divisors common to all the integers.   Excluding 1 though, what can be the LCD?

It is the smallest factor common to all the integers. A factor begins with a prim factor.

So it is the smallest of all prime factors common to all the integers.

Getting back to the example above, we have:

$576 = 2^6 3^2$ apples, $360 = 2^3 3^2 5$ pears, $180 = 2^2 3^2 5$ bananas, and $252 = 2^2 3^2 7$ oranges.

Then, 2 and 3 are all prime factors common to all the integers, and 2 is the smaller.
So the smallest factor common to all is 2, which can be therefore, the LCD.
So what is the LCD about?

Distributing objects equally in each kind, and minimizing the number of recipients, we want to find the LCD of the numbers of objects in all kinds. That is, distributing objects equally in each kind, and maximizing each number of objects in each kind each recipient gets, we want to find the LCD of the numbers of objects in all kinds.

- Next, what is LCM?

It is the acronym of Lest Common Multiple. So it is a multiple common and the smallest.

As in the case of GCD, when working with a set of objects, we often consider LCM, and the objects are usually integers, constants, monomials, or polynomials.

So for instance, finding an LCM integer, we work with a set of integers.

It is common to all the integers in the set. And also, since it is a multiple of all the integers, every integer in the set can divide it.   What then, is the next?

It is the smallest, too, so it is the least multiple every integer in the set can divide.

In other words, it is the least multiple common to all the integers in the set. And the same is true, too, for a set of constants, a set of monomials, or a set of polynomials.

Why do we need LCM, though?

Suppose we want to distribute objects evenly to a number of groups, in each of which though, the number of members can be different, but every member has to get the same amount. Suppose also, we want to minimize the amount each group gets.

For instance, one of three groups has 2 members, another group has 3, and the other has 5. In each group however, the amount every member gets is the same, and the total amount given to each group is the same, too. And also, we want to minimize the total amount each group gets.

Then, we want to find the least of all multiples common to all the numbers of members. The numbers are integers, of course, and the least is the LCM.

Suppose for instance, we want to distribute pencils to four schools, one of which has 576 students, another has 360, another has 180, and the other has 252, but want to allocate pencils evenly to all the schools, in each of which every student gets the same amount.

So since each school has a different number of students, the amount a student gets in one school is different from the amount a student gets in any of the other schools.

Then, how many pencils can each school get?

Every school gets the same amount, of course, which however, we want to find. So to begin with, suppose X is the number of pencils each school gets.

Then, since in each school, every student gets the same amount, the number of students in each school can divide X. So X is a multiple of the number of students in each school.

That is, X is a multiple common to the numbers of students in all the schools.
What particular number then, can X be?

Since each number of students is an integer, X is a multiple of an integer, and thus, can be infinitely large, so there is no maximum for it. There does exist the minimum, though. What then, is the minimum?

It is the minimum of X, of course.
To begin with, X is a multiple common to the numbers of students in all the schools.

So the minimum of X is the smallest of all multiples common to the numbers of students in all the schools, and the smallest is the LCM.   How can we find it, though?

Suppose A is a multiple of B, and is a multiple of C, too.
Then, A is a multiple common to B and C. Thus, B can divide A, and so can C.
So do we have to have: $A = BC$?

Not necessarily.
Suppose now, $B = uv$, and $C = vw$.

Then, u and v are prime factors of B, and v and w are prime factors of C.
Both of B and C can divide not only $uvvw = uv^2w$, but uvw, too.

Suppose next, $B = uv^2$, and $C = vw$.

Then again, u and v are prime factors of B, and v and w are prime factors of C.
Both of B and C can divide $u^2v^2w^2$, u^2v^2w, and uv^2w, but not uvw.

That's because $B = uv^2$ cannot divide uvw..

Suppose next, $B = uv^2$, and $C = vw^2$.

Then again, u and v are prime factors of B, and v and w are prime factors of C. Both of B and C can divide $u^2v^3w^3$, $u^2v^2w^2$, and uv^2w^2, but not uv^2w.

That's because $C = vw^2$ cannot divide uv^2w.

Suppose next, $B = uv^2w$, and $C = vw^2$.

Then, u, v, and w are prime factors of B, and v and w are prime factors of C. Both of B and C can divide $u^2v^3w^3$, $u^2v^2w^2$, and uv^2w^2, but not uv^2w.

That's because $C = vw^2$ cannot divide uv^2w.

Suppose next, $B = uv^2ws$, and $C = vw^2t$.

Then, u, v, w and s are prime factors of B, and v, w, and t are prime factors of C. Both of B and C can divide $u^2v^3w^3s^2t^2$, $u^2v^2w^2s^2t$, and uv^2w^2st, but not uv^2ws.

That's because $C = vw^2t$ cannot divide uv^2ws.

Therefore, a multiple common to a set of monomials has <u>at least</u> all prime factors all the monomials have, and the exponent to be used for each prime factor is <u>at least</u> the largest of all used for the prime factor.

Now, in the case of uv^2ws and vw^2t, all prime factors the two monomials uv^2ws and vw^2t have are u, v, w, s, and t, 1 is the largest exponent used for u, 2 is the one for each of v and w, and 1 is the one for each of s and t. So a multiple common to uv^2ws and vw^2t has to have as prime factors <u>at least</u> u, v, w, s, and t, and the exponents to be used for the prime factors are <u>at least</u> 1, 2, 2, 1, and 1 respectively in the order of u, v, w, s, and t.

Therefore, the multiple the <u>least</u> and common to a set of monomials has *all prime factors* all the monomials have, and the *exponent* to be used for each prime factor is *the largest* of all used for the prime factor. Thus, the LCM of uv^2ws and vw^2t is uv^2w^2st.

The same is true for integers and polynomials, too. So solving the problem with the pencil distribution, we want to begin with factorizing all the numbers of students.

Factorizing them all, we can see one of the four schools has $576 = 2^6 3^2$ students, another has $360 = 2^3 3^2 5$, another has $180 = 2^2 3^2 5$, and the other has $252 = 2^2 3^2 7$.

Therefore, all prime factors the product of all the numbers of students has are 2, 3, 5, and 7, and the largest exponents used for the prime factors are 6, 2, 1, and 1.

So the least common multiple of all the numbers of the students is $2^6 3^2 5 \cdot 7$, which is X, which is the minimum number of pencils each school gets.

Thus, since every school gets: $2^6 3^2 5 \cdot 7 = 20160$ pencils, we can see that:

Each student attending the school of $576 = 2^6 3^2$ students gets $5 \cdot 7 = 35$ pencils.
Each student attending the school of $360 = 2^3 3^2 5$ students gets $2^3 \cdot 7 = 56$ pencils.
Each student attending the school of $180 = 2^2 3^2 5$ students gets $2^4 \cdot 7 = 112$ pencils.
Each student attending the school of $252 = 2^2 3^2 7$ students gets $2^4 \cdot 5 = 80$ pencils.

That's not the only case where we can use LCM, of course.

We can use LCM when adding together fractions where denominators are different. Of course, the same is true for subtractions with such fractions, too, since subtractions are additions of the negatives. In such cases though, we often use just a common multiple.

Taking the LCM of all the denominators different, and using the LCM as the common denominator, we can add the fractions. So for instance, adding together $\frac{1}{2}$, $\frac{5}{3}$, and $\frac{7}{4}$, we can take the sum in such a way as follows.

Taking the LCM of 2, 3, and 4, which is 12, and using it as the common denominator of the fractions, we get: $\frac{1}{2} = \frac{6}{12}$, $\frac{5}{3} = \frac{20}{12}$, and $\frac{7}{4} = \frac{21}{12}$, so adding them up, we get $\frac{6+20+21}{12} = \frac{47}{12}$, which is the sum of $\frac{1}{2}$, $\frac{5}{3}$, and $\frac{7}{4}$.

Just taking a product of 2, 3, and 4 though, and using it as the common denominator, we can get the same result, too, of course, but calculation gets longer.

And the same is true, too, for fractions made of monomials and polynomials. So for instance, adding together $\frac{1}{2}$, $\frac{5a}{4c^2}$, and $\frac{2b(a+1)}{3c(a+b)}$, we can take the sum in a way as follows.

Taking the LCM of $\mathbf{2}$, $\mathbf{4c^2}$, and $\mathbf{3c(a + b)}$, and using it as the common denominator of the fractions, we get: $\frac{1}{2} = \frac{6c^2(a+b)}{12c^2(a+b)}$, $\frac{5a}{4c^2} = \frac{15a(a+b)}{12c^2(a+b)}$, and $\frac{2b(a+1)}{3(a+b)} = \frac{8bc^2(a+1)}{12c^2(a+b)}$, so adding them up, we get: $\frac{6c^2(a+b)+15a(a+b)+8bc^2(a+1)}{12c^2(a+b)}$, which is the sum of $\frac{1}{2}$, $\frac{5a}{4c^2}$, and $\frac{2b(a+1)}{3c(a+b)}$.

How come the LCM is: $\mathbf{12c^2(a + b)}$, though?

As other objects in math, LCM and GCD are tools, which are quite handy in many cases.

Doing algebra, we often use tools called GCD and LCM of polynomials.

And in the next section, we are going to look at how we can get such tools, and how they work.

GCD and LCM 2

Let's have a look at now, how we can get such tools, and how they work.

Then, we can see how $12c^2(a + b)$ can be the LCM of 2, $4c^2$, and $3c(a + b)$.

• We are going to begin with GCD of polynomials, though.

Suppose that $k = 0$, 1, 2, … n, where n is a nonnegative integer, and that Q_k and D are polynomials.

Suppose also, that for all k, all Q_k are different from each other.

Now, suppose first, that $M = DQ_0$.

Then, M is the product of D and Q_0, which are polynomials.
Thus, M is a polynomial, too, and in particular, is a multiple of D, so D is a divisor of M.
The same is true for Q_0, also.

Suppose next, $M_0 = DQ_0$, and $M_1 = DQ_1$.

Then, M_0 is a multiple of D, so D is a divisor of M_0.
And the same is true for M_1, too, so M_1 is a multiple of D, also.
So D is a divisor of M_1 as well as M_0.
Thus, D is a divisor common to M_0 and M_1.

Now, suppose further, that there is no factor common to Q_0 and Q_1.

That is, other than 1, there is no divisor common to Q_0 and Q_1.

In other words, Q_0 has no factor that can be a factor of Q_1, too, and vice versa.

For instance, if $Q_0 = (x + 1)(x + 2)$, and $Q_1 = (x + 3)(x + 4)$, all factors Q_0 has are $x + 1$ and $x + 2$, so neither of those factors can divide (that is, can be a factor of) Q_1.

Then, other than 1 and D, there is no divisor common to M_0 and M_1.
Then, D is not just a divisor common to M_0 and M_1.   What else then, is D?

D is the greatest divisor, too.
In other words, D is the greatest divisor common to M_0 and M_1.
That is, D is the GCD of M_0 and M_1.   How come?

The only divisors common to M_0 and M_1 are 1 and D.   Divisors of D can be 1 and itself.

A divisor that has more divisors is the bigger divisor.   So D is the GCD.

Suppose this time, $M_k = DQ_k$, where $k = 0, 1, 2, \ldots n$, where n is an integer ≥ 0.
That is, $M_0 = DQ_0, M_1 = DQ_1, M_2 = DQ_2, \ldots, M_n = DQ_n$.

Then, for each k, M_k is a multiple of D, so D is a divisor of M_k.
Thus, D is a divisor of every M_k for every k.
That is, D is a common divisor of all M_k for all k.

Now, suppose further, for a value of k, Q_k has no factor common to all the other Q_k.

For instance, Q_2 has no factor common to all the others, which are $Q_0, Q_1, Q_3, Q_4,\ldots$, and Q_n. (That is, no factor of Q_2 can be a factor common to $Q_0, Q_1, Q_3, Q_4, \ldots$, and Q_n.)
In other words, there is no factor common to all Q_k for all k.

Then, other than 1 and D, there is no divisor common to all M_k for all k.
Then, D is not just a divisor common to all M_k for all k.   What else, too, is it, then?

D is the greatest divisor, too.

In other words, **D** is the greatest divisor common to all M_k.

That is, **D** is the GCD of all M_k for all **k**.   How come?

The only divisors common to all M_k are 1 and **D**, and **D** has more divisors than 1 has.

A divisor that has more divisors is the bigger divisor. So **D** is the GCD.

• Let's now, take some examples more specific.

Suppose that $D = x + 1$, and that:

$Q_0 = (x + y)(y + 1)$

$Q_1 = (x + y)^2(y + 1)$

$Q_2 = (x + 2y)(y^2 + y + 4)$

$Q_3 = (x + 1)(x + y)(x + 3y)$.

Suppose also, that $M_k = DQ_k$, where $k = 0, 1, 2$, and **3**, that is:

$M_0 = DQ_0 = (x + 1)(x + y)(y + 1)$

$M_1 = DQ_1 = (x + 1)(x + y)^2(y + 1)$

$M_2 = DQ_2 = (x + 1)(x + 2y)(y^2 + y + 4)$

$M_3 = DQ_3 = (x + 1)(x + 1)(x + y)(x + 3y) = (x + 1)^2(x + y)(x + 3y)$.

Then, **D** is the greatest common divisor, that is, the GCD of all M_k for all **k**. How come?

To begin with, **D** is a divisor of every M_k for every **k**.

So **D** is a divisor common to all M_k for $k = 0, 1, 2$, and **3**.

Next, there is no factor common to Q_0, Q_1, Q_2, and Q_3.

That is, for instance, all factors of Q_0 are $(x + y)$ and $(y + 1)$, and neither of all the factors can divide at the same time all of Q_1, Q_2, and Q_3.

And the same is true for each of Q_1, Q_2, and Q_3, too.
So other than 1 and D, there is no divisor common to all the Ms.

Next, D has more divisors than 1 has, and therefore, is the GCD.

Suppose this time, that $D = (x + 1)(x + 2)$, and that:

$Q_0 = (x + y + z)(y + 1)$
$Q_1 = (x + z)(y + 1)^2$
$Q_2 = (x + y)(y^2 + x)$
$Q_3 = (x + 1)(x + y)$
$Q_4 = (x + 1)(x + 2)(x + y)(y + 1)$.

Suppose also, $M_k = DQ_k$, where $k = 0, 1, 2, 3$, and 4, that is:

$M_0 = DQ_0 = (x + 1)(x + 2)(x + y + z)(y + 1)$
$M_1 = DQ_1 = (x + 1)(x + 2)(x + z)(y + 1)^2$
$M_2 = DQ_2 = (x + 1)(x + 2)(x + y)(y^2 + x)$
$M_3 = DQ_3 = (x + 1)(x + 2)(x + 1)(x + y) = (x + 1)^2(x + 2)(x + y)$
$M_4 = DQ_4 = (x + 1)(x + 2)(x + 1)(x + 2)(x + y)(y + 1) = (x + 1)^2(x + 2)^2(x + y)(y + 1)$.

Then, D is the greatest common divisor, that is, the GCD of all M_k for all k. How come?

To begin with, D is a divisor of every M_k for every k.

So D is a divisor common to all M_k for all k.

Next, $(x + 1)$ is a factor of D, yet it can divide all M_k for all k, too.

The same is true for $(x + 2)$, too.

So $(x + 1)$ and $(x + 2)$, too, are divisors common to all M_k for all k.

Next, there is no factor common to all Qs, which are Q_0, Q_1, Q_2, Q_3, and Q_4.

That is, for instance, all factors of Q_3 are $(x + 1)$ and $(x + y)$, and neither of all the factors can be common to all of Q_0, Q_1, Q_2, and Q_4.

And the same is true for each of Q_0, Q_1, Q_2, and Q_4, too.

So other than 1, $(x + 1)$, $(x + 2)$, and $(x + 1)(x + 2) = D$, there is no divisor common to all the Ms.

Now, divisors of $x + 2$ is 1 and itself, and the same is true for $x + 1$, too, but those of D is $x + 1$ and $x + 2$, together with 1 and itself. The more divisors, the bigger divisor.

Among all the common divisors, D has the most number of divisors, so D is the GCD.

Suppose for another example, that:

$$M_0 = 144(x + 1)^2(x + y)^2(y + 1) = 2^4 3^2(x + 1)^2(x + y)^2(y + 1)$$
$$M_1 = -216(x + 1)(x + y)^3(y + 2) = -2^3 3^3(x + 1)(x + y)^3(y + 2)$$
$$M_2 = 360(x + 1)(x + y)^2(y + 3)^2 = 2^3 3^2 5(x + 1)(x + y)^2(y + 3)^2$$
$$M_3 = 504(x + 1)(x + y)^2(y^2 + y + 4) = 2^3 3^2 7(x + 1)(x + y)^2(y^2 + y + 4)$$
$$M_4 = -792(x + 1)^2(x + y)^3(2y + 5x) = -2^3 3^2 11(x + 1)^2(x + y)^3(2y + 5x)$$

Then, we can begin with finding divisors common to all M_k for $k = 0$, 1, 2, 3, and 4. For instance: $x + 1$, $2(x + 1)$, $3(x + 1)$, etc. are common divisors. Looking for GCD though, we may not want to begin with just finding all divisors common.   Why not?

There are quite a few, or rather, excessively many divisors common to all M_k. What then, do we need to begin with?

For instance, the GCD of a set of integers is normally a product of factors common to all the integers. So we want to begin with factorizing all the integers. And the same is true for monomials and polynomials, too.   What are the factors, though?

Suppose all the integers are now fully factorized, and two prime factors are common to all the integers. Then, the GCD is the product of two factors common to all the integers.

One of the two factors is a power where the base is one of the two prime factors, and the exponent is the smallest of all used for the prime factor.

The other is a power, too, where the base is the other prime factor, and the exponent is the smallest of all used for the other prime factor.

So the product of the two powers above is the greatest of all divisors common to all the integers, and thus, is the GCD.   How come?

Of all powers where the base is a prime factor common to all the integers, what power is the one the greatest and common to all the integers?

It is the one where the exponent is the smallest of all used for the prime factor.

Therefore, of all integers, the product of two factors, which are the two powers above, is the divisor the greatest and common to all the integers, so the product is the GCD.

And the same is true for monomials and polynomials, too.

- Let's find for instance, the GCD of 360, 144, and 504.

To begin with, we want to factorize all the integers we want to find the GCD of.

Then, we get: $360 = 2^3 3^2 5$, $432 = 2^4 3^3$, and $504 = 2^3 3^2 7$.

Next, finding all prime factors common to all the integers, we get 2 and 3.

Next, we find the smallest of all exponents used for each of the prime factors above.
The smallest exponent used for the prime factor 2 is 3, and for the prime factor 3, 2 is the smallest exponent used. Then, we get two powers 2^3 and 3^2.

In the power 2^3, the base is 2, which is a prime factor common to all the integers, and the exponent is 3, which is the smallest of all used for the prime factor 2.

So among all powers where the base is 2, 2^3 is the divisor the greatest and common to all the integers $360 = 2^3 3^2 5$, $432 = 2^4 3^3$, and $504 = 2^3 3^2 7$.

In the power 3^2, the base is 3, which is a prime factor common to all the integers, and the exponent is 2, which is the smallest of all used for the prime factor 3.

So among all powers where the base is 3, 3^2 is the divisor the greatest and common to all the integers 360, 432, and 504.

Thus, two powers 2^3 and 3^2 are all the factors we take the product of. That is, among all integers, the integer $2^3 3^2$, which is 72 is the greatest of all common to 360, 432, and 504, and therefore, is the GCD.

So in short, finding the GCD of a set of integers, we get all the prime factors common to all the integers and the exponents the smallest of all used for the prime factors, and then, take the product of all the powers. Then, the product is the GCD.

And the same is true for GCD of monomials and polynomials, too.

So finding the GCD of a set of polynomials, too, we want to do it *prime factor by prime factor*, and thus, we want to factorize all the polynomials first.

Now, getting back to the example, we have:

$$M_0 = 2^4 3^2 (x + 1)^2 (x + y)^2 (y + 1), \quad M_1 = -2^3 3^3 (x + 1)(x + y)^3 (y + 2),$$
$$M_2 = 2^3 3^2 5(x + 1)(x + y)^2 (y + 3)^2, \quad M_3 = 2^3 3^2 7(x + 1)(x + y)^2 (y^2 + y + 4), \text{ and}$$
$$M_4 = -2^3 3^2 11(x + 1)^2 (x + y)^3 (2y + 5x).$$

• To begin with, we factorize all the polynomials we want to find GCD of.
All the polynomials M_k are all fully factorized already, so we are ready to get the GCD.

• Next, finding all the prime factors common to all M_k, we get: **2, 3, $x + 1$,** and $x + y$.

- Next, we find the smallest of all exponents used for each of the prime factors. Then:

3 is the smallest exponent used for the prime factor 2.
2 is the smallest exponent used for the prime factor 3.
1 is the smallest exponent used for the prime factor $x + 1$.
2 is the smallest exponent used for the prime factor $x + y$.

- Next, we get the powers common to all the polynomials M_k.

Then, we get four powers 2^3, 3^2, $(x + 1)^1 = x + 1$, and $(x + y)^2$, each of which is a factor common to all the polynomials. In the power $(x + 1)^1$, the base is $x + 1$, which is one of the prime factors common to all the polynomials, and the exponent is 1, which is the smallest of all used for the prime factor $x + 1$.

- Next, we get the product of the four factors, which are the four powers above.

Then, we simply get: $2^3 3^2 (x + 1)(x + y)^2$, which is the GCD of all the polynomials M_k.

We have some negatives among M_k though, so why <u>not</u> GCD = $-2^3 3^2 (x + 1)(x + y)^2$?

That's because we normally take a positive integer for a divisor.

The same is true for divisors in constant, monomial, and polynomial, too.
Of course, we can put the GCD above this way, too: $72(x + 1)(x + y)^2$.

Usually though, finding the GCD, we do it in such a way as follows:

- To begin with, we factorize all the polynomials we want to find GCD of.
Using the example above, we get:

$2^4 3^2 (x + 1)^2 (x + y)^2 (y + 1)$, $-2^3 3^3 (x + 1)(x + y)^3 (y + 2)$, $2^3 3^2 5(x + 1)(x + y)^2 (y + 3)^2$,
$2^3 3^2 7(x + 1)(x + y)^2 (y^2 + y + 4)$,   and $-2^3 3^2 11(x + 1)^2 (x + y)^3 (2y + 5x)$.

- Next, we find all prime factors common to all the polynomials.

Then, we get: **2, 3, $x + 1$**, and $x + y$.

• Next, we take the product of all those factors above:

Then, we get: $2 \cdot 3(x + 1)(x + y)$.

• Next, we apply the smallest of all exponents used for each of the prime factors above.

Then, we get: $2^3 3^2 (x + 1)(x + y)^2$, which is the GCD.

Now, let's for another example, find the GCD of polynomials as follows:

$M_0 = 144a^3b^2xy(x + 1)^2(x + y)^2(y + 1) = 2^4 3^2 a^3 b^2 xy(x + 1)^2(x + y)^2(y + 1)$

$M_1 = -216a^2by^2(x + 1)(x + y)^3(y + 2) = -2^3 3^3 a^2 by^2(x + 1)(x + y)^3(y + 2)$

$M_2 = 360a^2bxy(x + 1)(x + y)^2(y + 3)^2 = 2^3 3^2 5a^2 bxy(x + 1)(x + y)^2(y + 3)^2$

$M_3 = 504a^3b^2cy^3(x + 1)(x + y)^2(y^2 + y + 4) = 2^3 3^2 7a^3 b^2 cy^3(x + 1)(x + y)^2(y^2 + y + 4)$

$M_4 = -792a^4bc^2y(x + 1)^2(x + y)^3(2y + 5x) = -2^3 3^2 11a^4 bc^2 y(x + 1)^2(x + y)^3(2y + 5x)$.

• To begin with, we factorize all the polynomials we want to find GCD of.

All the polynomials M_k are all fully factorized already, so we are ready to get the GCD.

• Next, finding all prime factors common to all M_k, we get: **2, 3, a, b, y, $x + 1$**, and $x + y$.

• Next, taking the product of all those factors above, we get: $2 \cdot 3aby(x + 1)(x + y)$.

• Next, we apply the smallest of all exponents used for each of the prime factors above.

Then, we get: $2^3 3^2 a^2 by(x + 1)(x + y)^2$, which is the GCD.

So in short, finding GCD, what do we do?

Put in a form of a product all the prim factors common, and then, apply to each of the factors the smallest exponent used.

• Now, let's move on to LCM of polynomials.

LCM has quite the opposite sense when it is compared to GCD.

Whereas GCD is a divisor the greatest, LCM is a multiple the smallest.

Both share the same nature though. Both are common to many integers or are common to many polynomials.

So GCD is the common divisor the greatest, LCM is the common multiple the smallest.

Now, suppose first, B and C are monomials different from each other.
Suppose next, A is a multiple of B, and also, is a multiple of C.

Then, A is a multiple common to B and C. So do we have to have $A = BC$?

Not necessarily, of course.   Why not though?

Suppose first, B and C are prime to each other.
That is, B does not have any prime factor that C has. For instance, $B = uv$, and $C = st$.

Then, A has to have as prime factors u, v, s, and t <u>at least</u>.

So A can have more prime factors, too.

That's because A is a multiple common to B and C, so B can divide A, and so can C.
For instance, A can be $uvst = BC$, $2uv^2st = 2vBC$, $3cuvst = 3cBC$, $uvstxy = BCxy$, etc.

Suppose next, B has some prime factors that C has.

Then, it can rather be the case where BC can be a multiple of A.   How come?

Suppose for instance, $B = uv$, and $C = vw$.

Then, A has to have u, v, and w as prime factors <u>at least</u>, since B and C both divide A, because A is a multiple common to B and C.

So A can be uvw, because B divides A, and so does C, since A is a multiple common to B and C.
And we have: $BC = uv^2w$. So if $A = uvw$, we get: $BC = vA$, so BC is a multiple of A.

And of course, in this case, too, A can have more prime factors other than u, v, and w, because A is a multiple common to B and C.

For instance, A can be uvw, $2uvw$, au^2vw, etc.   Then, B divides A, and so does C.

So there can be infinitely many monomials that can be A, a multiple common to B and C.

And thus, there does not exist the greatest multiple common if no other condition is applied.   There does exist though, the least common multiple, called LCM.

So let's see now, how it can be made, that is, the principle behind the LCM.

In the case of the examples A, B, and C above, we have found that:

• A multiple common to a set of monomials has to have <u>at least</u> all the prime factors all the monomials have, and the exponent to be used for each prime factor is <u>at least</u> the largest of all used for the prime factor.

So let's now, for instance, consider the case of a multiple common to uv^2ws and vw^2t.

First, all the prime factors the two monomials uv^2ws and vw^2t have are: u, v, w, s, and t.

Next, 1 is the largest exponent used for u, 2 is the one for each of v and w, and 1 is the one for each of s and t.

So a multiple common to uv^2ws and vw^2t has to have as prime factors <u>at least</u> u, v, w, s, and t, and the exponents to be used for the prime factors are <u>at least</u> 1, 2, 2, 1, and 1 respectively in the order of u, v, w, s, and t.

Therefore, the multiple the <u>least</u> and common to the two monomials has *all the prime factors* all the two monomials have, and the *exponent* to be used for each prime factor is *the largest* of all used for the prime factor. Thus, the LCM of uv^2ws and vw^2t is: uv^2w^2st.

And the same is true for integers and polynomials, too.

So in the case of polynomials, too, the multiple the <u>least</u> and common to a set of polynomials, that is, the LCM has *all the prime factors* all the polynomials have, and the *exponent* to be used for each prime factor is *the largest* of all used for the prime factor.

Let's now, for instance find the LCM of polynomials below:

$P_0 = (x + y)(y + 1)$, $\quad P_1 = (x + y)^2(y + 1)$,

$P_2 = (x + 2y)(y^2 + y + 4)$, $\quad$ and $P_3 = (x + 1)(x + y)(x + 3y)$.

Then, all the polynomials are fully factorized already, so we are ready to find the LCM.

• So next, we just take the product of <u>all the prime factors</u> that all the polynomials have.

Then, we get: $(x + y)(y + 1)(x + 2y)(y^2 + y + 4)(x + 1)(x + 3y)$.

• Next, apply to each prime factor above <u>the largest exponent</u> of all used for the prime factor.

Then, 2 is the largest used for the factor $(x + y)$, and 1 is the largest used for each of the factors $(y + 1)$, $\quad (x + 2y)$, $\quad (y^2 + y + 4)$, $\quad (x + 1)$, $\quad$ and $(x + 3y)$.

Therefore, the LCM is: $(x + y)^2(y + 1)(x + 2y)(y^2 + y + 4)(x + 1)(x + 3y)$.

Suppose this time, we want to find the LCM of polynomials below:

$P_0 = (x + 1)(x + y)(y + 1)$

$P_1 = (x + 1)(x + y)^2(y + 1)$

$P_2 = (x + 1)(x + 2y)(y^2 + y + 4)$

$P_3 = (x + 1)^2(x + y)(x + 3y)$

Then again, all the polynomials are factorized already, so we are ready to find the LCM.

• So next, we just take the product of <u>all the prime factors</u> all the polynomials have.

Then, we get: $(x + 1)(x + y)(y + 1)(x + 2y)(y^2 + y + 4)(x + 3y)$.

That's because the LCM has all the prime factors all the polynomials have so that each and every polynomial can divide the LCM.

• Next, apply to each prime factor above <u>the largest exponent</u> of all used for the prime factor.

Then, 2 is the largest for each of $(x + 1)$ and $(x + y)$, and 1 is the largest used for each of $(y + 1)$, $(x + 2y)$, $(y^2 + y + 4)$, and $(x + 3y)$.

Therefore, the LCM is: $(x + 1)^2(x + y)^2(y + 1)(x + 2y)(y^2 + y + 4)(x + 3y)$.

Let's next, find the LCM of polynomials below:

$$P_0 = 144(x + 1)^2(x + y)^2(y + 1) = 2^4 3^2(x + 1)^2(x + y)^2(y + 1)$$

$$P_1 = -216(x + 1)(x + y)^3 = -2^3 3^3(x + 1)(x + y)^3$$

$$P_2 = 360(x + 1)(x + y)^2(y + 3)^2 = 2^3 3^2 5(x + 1)(x + y)^2(y + 3)^2$$

$$P_3 = 504(x + 1)(x + y)^2 = 2^3 3^2 7(x + 1)(x + y)^2$$

$$P_4 = -792(x + 1)^2(x + y)^3(2y + 5x) = -2^3 3^2 11(x + 1)^2(x + y)^3(2y + 5x).$$

Then, all the polynomials are fully factorized already, so we are ready to find the LCM.

• So next, we want to take the product of <u>all the prime factors</u> all the polynomials have.

Then, we can simply get: $2\cdot3\cdot5\cdot7\cdot11(x + 1)(x + y)(y + 1)(y + 3)(2y + 5x)$.

• Next, we want to apply to each prime factor above <u>the largest exponent</u> of all used for the prime factor.

Then, 4 is the largest used for the prime factor 2, 3 is the one used for the factor 3, 1 is the one used for the factor 5, 1 is used for 7, 1 is used for 11, 2 is used for $(x + 1)$, 3 is used for $(x + y)$, 1 is used for $(y + 1)$, 2 is used for $(y + 3)$, and 1 is used for $(2y + 5x)$.

Therefore, the LCM is: $2^4 3^3 5 \cdot 7 \cdot 11(x + 1)^2(x + y)^3(y + 1)(y + 3)^2(2y + 5x)$.

We can of course, put the LCM above this way, too:

$166320(x + 1)^2(x + y)^3(y + 1)(y + 3)^2(2y + 5x)$ since $166320 = 2^4 3^3 5 \cdot 7 \cdot 11$.

Now, let's for another example, find the LCM of polynomials as follows:

$$M_0 = 144a^3b^2xy(x + 1)^2(x + y)^2(y + 1) = 2^4 3^2 a^3 b^2 xy(x + 1)^2(x + y)^2(y + 1)$$

$$M_1 = -216a^2by^2(x + 1)(x + y)^3 = -2^3 3^3 a^2 by^2(x + 1)(x + y)^3$$

$$M_2 = 360a^2bxy(x + 1)(x + y)^2(y + 3)^2 = 2^3 3^2 5a^2 bxy(x + 1)(x + y)^2(y + 3)^2$$

$$M_3 = 648a^3b^2cy^3(x + 1)(x + y)^2 = 2^3 3^4 a^3 b^2 cy^3(x + 1)(x + y)^2$$

$$M_4 = -72a^4bc^2y(x + 1)^2(x + y)^3 = -2^3 3^2 a^4 bc^2 y(x + 1)^2(x + y)^3.$$

Then, all the polynomials are fully factorized already, so we are ready to find the LCM.

• So next, we just take the product of <u>all the prime factors</u> all the polynomials have.

Then, we can simply get: $2 \cdot 3 \cdot 5 \cdot 7 \cdot abcxy(x + 1)(x + y)(y + 1)(y + 3)$.

• Next, we want to apply to each prime factor <u>the largest exponent</u> of all used for the prime factor.

Then, 4 is the largest used for the prime factor 2, 4 is the one used for the factor 3, 1 is the one used for the factor 5, 4 is used for the factor a, 2 is used for b, 2 is used for c, 1 is used for x, 3 is used for y, 2 is used for $(x + 1)$, 3 is used for $(x + y)$, 1 is used for $(y + 1)$, and 2 is used for $(y + 3)$.

Therefore, the LCM is: $2^4 3^4 5a^4 b^2 c^2 xy^3(x + 1)^2(x + y)^3(y + 1)(y + 3)^2$.

We can put the LCM above this way, too: $6480a^4 b^2 c^2 xy^3(x + 1)^2(x + y)^3(y + 1)(y + 3)^2$, since $6480 = 2^4 3^4 5$.

Examples 1 in GCD and LCM

Find the GCD (Greatest Common Divisor) and LCM (Least Common Multiple) in each of the cases below.

0. $72x^2y^3z^5$ and $16x^2y^4z^7$

1. $x^2(x+y)^4(x+z)$ and $x^5(x+y)^2(x+z)^8$

2. 72 and $16x^2y^4z^7$

Suggestions or Solutions
To the **Problem** in the Example **0**

Find the GCD and LCM of $72x^2y^3z^5$ and $16x^2y^4z^7$.

To begin with, what is GCD?

It is the acronym of Greatest Common Divisor. So it is a divisor common and the largest. Looking for GCD though, we don't want to just begin with finding all divisors common. What then, do we begin with?

Suppose for instance, we want to find the GCD of a set of integers.

Then usually, we take a product of all factors, each of which is a power where the base is a prime factor common to all the integers, and the exponent is the smallest of all used for the prime factor. So the product has all the prime factors common to all the integers.

So the product is the divisor the greatest and common to all the integers, and thus, is the GCD. Thus, we want to begin with factorizing all the integers.

And the same is true for monomials and polynomials, too.

Let's find for instance, the GCD of 216, 144, and 360.

- First, factorizing all the integers, we get: $216 = 2^3 3^3$, $144 = 2^4 3^2$, and $360 = 2^3 3^2 5$.

- Next, finding all the prime factors common to all the integers, we get 2 and 3.

- Next, we find the smallest of all exponents used for each of the prime factors above.

Then, 3 is the smallest exponent used for the prime factor 2, and for the prime factor 3, 2 is the smallest exponent used.

So we get two powers, in one of which, the base is 2, and the exponent is 3, and in the other, the base is 3, and the exponent is 2. That is, we get 2^3 and 3^2.

Thus, taking the product of two powers, we get the GCD, which is $2^2 3^2$.

> So in short, finding the GCD of a set of integers, we get all the prime factors common to all the integers, together with the exponents the smallest of all used for the prime factors, and then, take the product of the powers.

And the same is true for a GCD of monomials and polynomials, too.

Now in this problem, we want to find the GCD of two monomials $72x^2y^3z^5$ and $16x^2y^4z^7$.

- To begin with, factorizing the monomials, we get: $2^3 3^2 x^2 y^3 z^5$ and $2^4 x^2 y^4 z^7$.

- Next, finding all the prime factors common to both monomials, we get: $2, x, y,$ and z.

- Next, apply to each factor above the smallest of all exponents used for the factor.

3 is the smallest exponent used for the prime factor 2.
2 is the smallest exponent used for the prime factor x.
3 is the smallest exponent used for the prime factor y.
5 is the smallest exponent used for the prime factor z.

- So we get four powers, which are $2^3, x^2, y^3,$ and z^5.

For instance, in the power x^2, the base is x, which is one of the prime factors common to both monomials, and the exponent is 2, which is the smallest of all the exponents used for the prime factor x.

Thus, taking the product of all the powers above, we get: $2^3 x^2 y^3 z^5$, which is the GCD of the two monomials $72x^2y^3z^5$ and $16x^2y^4z^7$.

•• Let's next, move on to the LCM. Then, to begin with, what do we mean by LCM?

It is the acronym of Lest Common Multiple. So it is a multiple common and the smallest.

Suppose for instance, we want to find the LCM of a set of integers.
Suppose also, all the integers are fully factorized already.

Then, a multiple common to all the integers is a product of powers, has <u>at least</u> all the prime factors all the integers have, and the exponent to be used for each prime factor is <u>at least</u> the largest of all the exponents used for the prime factor.

Thus, the LCM, the <u>least</u> of all the multiples common to a set of integers has *all the prime factors* all the integers have, and the *exponent* to be used for each prime factor is *the largest* of all the exponents used for the prime factor.

●●● So in short, finding the LCM of a set of integers, we get all the prime factors all the integers have, together with the exponents the largest of all used for the prime factors, and then, take the product of the powers.

And the same is true for monomials and polynomials, too.

Now, in the case of $72x^2y^3z^5$ and $16x^2y^4z^7$, we can factorize them to $2^33^2x^2y^3z^5$ and $2^4x^2y^4z^7$.

So a multiple common to both monomials has to have as prime factors <u>at least</u> $2, 3, x, y$, and z, and the exponents to be used for the prime factors are <u>at least</u> 4, 2, 2, 4, and 7 respectively in the order of $2, 3, x, y$, and z.

●●● So in short, finding the LCM of a set of monomials, we get all the prime factors all the monomials have, together with the exponents the largest of all used for the prime factors, and then, take the product of the powers.

So first, finding all the prime factors the two monomials have, we get: $2, 3, x, y$, and z.

Next, 4 is the largest exponent used for the factor 2, 2 is the one used for the factor 3, 2 is the one used for x, 4 is the one used for y, and 7 is the one for z.

Therefore, the LCM of $2^33^2x^2y^3z^5$ and $2^4x^2y^4z^7$ is: $2^43^2x^2y^4z^7$.

Suggestions or Solutions
To the **Problem** in the Example **1**

Find the GCD and LCM of $x^2(x+y)^4(x+z)$ and $x^5(x+y)^2(x+z)^8$.

A GCD is a divisor common and the largest.

Thus, the GCD of a set of polynomials is the greatest of all the divisors common to all the polynomials. More specifically, it is the product of all factors common to all the polynomials.   What are those factors, though?

The factors are powers, in each of which the base is a prime factor common to all the polynomials, and the exponent is the smallest of all used for the prime factor.

So the product has all the prime factors common to all the polynomial, and thus, is the divisor the greatest and common to all the polynomials, that is, the GCD.

Let's now, begin with the GCD of $12x^2(x+y)^4(x+z)$ and $8x^5(x+y)^2(x+z)^8$.

• First, factorizing the two polynomials, we get:
$2^2 3x^2(x+y)^4(x+z)$ and $2^3 x^5(x+y)^2(x+z)^8$.

• So next, all the prime factors common to both polynomials are: $2, x, x+y$, and $x+z$.

• Next, find the smallest of all exponents used for each of the factors above.

Then, 2 is the smallest exponent used for each of the three factors $2, x$, and $x+y$, and for the factor $x+z$, 1 is the smallest exponent used.

• So four powers $2^2, x^2, (x+y)^2$, and $(x+z)^1$ are factors common to $2^2 3x^2(x+y)^4(x+z)$ and $2^3 x^5(x+y)^2(x+z)^8$.

Note that $x + z = (x + z)^1$, so taking $(x + z)$ as a power, we take as the base the binomial $(x + y)$, which is one of the prime factors common to both polynomials, and we take 1 as the exponent. And 1 is the smallest of all the exponents used for the prime factor $x + y$.

• And next, take the product of all the four powers above.

Then, we get: $2^2 x^2 (x + y)^2 (x + z)$, which is the GCD.   How come though?

First, each of the two polynomials $2^2 3^1 x^2 (x + y)^4 (x + z)$ and $2^3 x^5 (x + y)^2 (x + z)^8$ is a product of powers, in each of which the base is: 2, 3, x, $x + y$, or $x + z$.

And next:

Of all powers of 2, the one where the exponent is 2 can divide both polynomials, and is the greatest.

Of all powers of x, the one where the exponent is 2 can divide both polynomials, and is the greatest.

Of all powers where the base is $x + y$, the one where the exponent is 2 can divide both polynomials, and is the greatest.

Of all powers of $(x + z)$, the one where the exponent is 1 can divide both polynomials, and is the greatest.

Therefore, the product of all the four powers, that is, $2^2 x^2 (x + y)^2 (x + z)$ is the greatest divisor common to both polynomials, and is called the GCD.

Usually though, we find the GCD the way below, which is in fact, no other than the way above though.

• First, factorize all the polynomials: $12 x^2 (x + y)^4 (x + z)$ and $8 x^5 (x + y)^2 (x + z)^8$.

Then, we get: $2^2 3 x^2 (x + y)^4 (x + z)$ and $2^3 x^5 (x + y)^2 (x + z)^8$.

• Next, find all the prime factors common to all the polynomials.

Then, we get **2**, x, $x + y$, and $x + z$.

• Next, take the product of all the prime factors above, and then, apply the smallest of all exponents used for each of the prime factors.

Then, we get: $2^2x^2(x + y)^2(x + z)$, which is the GCD.

●●● Therefore in short, finding the GCD of a set of polynomials, we get all the prime factors common to all the polynomials, together with the exponents the smallest of all used for the prime factors, and then, take the product of the powers.

And the same is true for integers and monomials, too.

• Let's next, move on to the LCM.

The LCM is the least common multiple.
So the LCM of a set of polynomial is the smallest of all multiples common to all the polynomials in the set.   How then, can we get the LCM?

The LCM has to be a common multiple, first, and next, has to be the least.
So beginning with a common multiple, we can put it the way below:

A common multiple of a set of polynomials has to have <u>at least</u> all the prime factors all the polynomials have, because each of all the polynomials has to divide the common multiple.
And the exponent to be used for each prime factor is <u>at least</u> the largest of all the exponents used for the prime factor.

So next, the <u>least</u> of all the multiples common to all the polynomials has to have *all the prime factors* all the polynomials have, and the *exponent* to be used for each prime factor is *the largest* of all the exponents used for the prime factor.

And the same is true for integers and monomials, too.

Now, in this example, we want to get the LCM of two polynomials, which are below:

$12x^2(x + y)^4(x + z)$ and $8x^5(x + y)^2(x + z)^8$, which are factorized to $2^23x^2(x + y)^4(x + z)$ and $2^3x^5(x + y)^2(x + z)^8$.

So all the prime factors both polynomials have are: $2, 3, x, x + y$, and $x + z$.

And next:

3 is the largest exponent used for the factor 2,

1 is the largest exponent used for the factor 3,

5 is the largest exponent used for x,

4 is the largest exponent used for $x + y$,

and 8 is the largest exponent used for $x + z$.

So a multiple common to both polynomials has as prime factors <u>at least</u> $2, 3, x, x + y$, and $x + z$. And the exponents to be used for the prime factors are <u>at least</u> 3, 1, 5, 4, and 8 respectively in the order of $2, 3, x, x + y$, and $x + z$.

And thus, the LCM, the <u>least</u> of all the multiples common to both polynomials is:

$2^33x^5(x + y)^4(x + z)^8$.

Suggestions or Solutions
To the **Problem** in the Example **2**

Find the GCD and LCM of 72 and $16x^2y^4z^7$.

To begin with, a GCD is a divisor common and the largest.

So the GCD of **72** and **$16x^2y^4z^7$** is the greatest of all divisors common to **72** and **$16x^2y^4z^7$**.

And getting the GCD, we get the product of all powers, in each of which the base is a prime factor common to both **72** and **$16x^2y^4z^7$**, and the exponent is the smallest of all used for the prime factor.

- So we want to begin with factorizing **72** and **$16x^2y^4z^7$**. Then, we get **$2^3 3^2$** and **$2^4 x^2 y^4 z^7$**.

- Next, finding all the prime factors common to **$2^3 3^2$** and **$2^4 x^2 y^4 z^7$**, we get **2** only.

So in this case, we get one power only where the base is a prime factor common to **$2^3 3^2$** and **$2^4 x^2 y^4 z^7$**, and therefore, we don't need to take such a product as described above.

Thus, the GCD is just a power, where the base is **2**, which is the prime factor, and the exponent is the smallest of all the exponents used for the prime factor **2**.

The smallest exponent is 3. So the GCD is **2^3**.

Let's next, move on to the LCM, which is the smallest and common multiple.

To begin with, for instance, a multiple common to a set of integers has <u>at least</u> all the prime factors all the integers have, and the exponent to be used for each prime factor is <u>at least</u> the largest of all used for the prime factor.

So the LCM, the multiple the <u>least</u> and common to all the integers has to have *all the prime factors* all the integers have, and the *exponent* to be used for each prime factor is *the largest* of all used for the prime factor.

Now, in the case of **72** and $16x^2y^4z^7$, which are fully factorized to 2^33^2 and $2^4x^2y^4z^7$, all the prime factors used in both are: **2, 3**, x, y, and z.

4 is the largest exponent used for the factor 2,

2 is the largest exponent used for 3,

2 is the largest exponent used for x,

4 is the largest exponent used for y,

and 7 is the largest exponent used for z.

So a multiple common to 2^33^2 and $2^4x^2y^4z^7$ has as prime factors <u>at least</u> 2, 3, x, y, and z, and the exponents to be used for the prime factors are <u>at least</u> 4, 2, 2, 4, and 7 respectively.

Therefore, the LCM of 2^33^2 and $2^4x^2y^4z^7$ is: $2^43^2x^2y^4z^7$.

Examples 2 in GCD and LCM

Find the GCD (Greatest Common Divisor) and LCM (Least Common Multiple) in each of the cases below.

0. $16x^2y^4z^7$ and $x^5(x+y)^2(x+z)^8$

1. 81, $16y^4z^7$, and $x^5(x+y)^2(x+z)^8$

2. x, y, and z

3. x^2+3x+2 and x^2-1

4. $(x+1)^2(x+2)(x+3)^2$, $x(x-1)(x+1)^4(x+3)^3$,   and $(x+2)^3(x+1)x^3(x+3)^3$

Suggestions or Solutions
To the **Problem** in the Example **0**

Find the GCD and LCM of $16x^2y^4z^7$ and $x^5(x+y)^2(x+z)^8$.

- A GCD is a divisor common and the largest.

So the GCD of **$16x^2y^4z^7$** and $x^5(x+y)^2(x+z)^8$ is the greatest of all divisors common to **$16x^2y^4z^7$** and $x^5(x+y)^2(x+z)^8$.

And getting the GCD, we get the product of all powers, in each of which the base is a prime factor common to both expressions above, and the exponent is the smallest of all used for the prime factor.

- So beginning with getting **$16x^2y^4z^7$** fully factorized, we get: **$2^4x^2y^4z^7$**.

- Next, finding all the prime factors common to **$2^4x^2y^4z^7$** and $x^5(x+y)^2(x+z)^8$, we get x only.

Thus, in this case, too, we get one power only where the base is a prime factor common to **$2^4x^2y^4z^7$** and $x^5(x+y)^2(x+z)^8$, so we don't need to take such a product as stated above.

Thus, the GCD in this case is just a power, where the base is x, which is the prime factor, and the exponent is the smallest of all exponents used for the prime factor x.

The smallest exponent is 2. So the GCD is: x^2.

Let's next, move on to the LCM, which is a multiple common and the smallest.

To begin with, for instance, a common multiple of a set of integers has <u>at least</u> all the prime factors all the integers have, and the exponent to be used for each prime factor is <u>at least</u> the largest of all used for the prime factor.

So the LCM, the multiple the <u>least</u> and common to all the integers has *all prime factors* all the integers have, and the *exponent* to be used for each prime factor is *the largest* of all used for the prime factor.

Now, in the case of $x^5(x + y)^2(x + z)^8$ and $16x^2y^4z^7$, which is factorized to $2^4x^2y^4z^7$, all the prime factors used in both are $2, x, y, z, x + y$, and $x + z$.

4 is the largest exponent used for the factor 2,

5 is the largest exponent used for the factor x,

4 is the largest exponent used for y,

7 is the largest exponent used for z,

2 is the largest exponent used for $x + y$,

and 8 is the largest exponent used for $x + z$.

So a multiple common to $x^5(x + y)^2(x + z)^8$ and $2^4x^2y^4z^7$ has as prime factors <u>at least</u> $2, x, y, z, x + y$, and $x + z$, and the exponents to be used for the prime factors are <u>at least</u> 4, 5, 4, 7, 2 and 8 respectively.

Therefore, the LCM is: $2^4x^5y^4z^7(x + y)^2(x + z)^8$.

Suggestions or Solutions

Find the GCD and LCM of 81, $16y^4z^7$, and $x^5(x+y)^2(x+z)^8$.

GCD is a divisor common and the largest.

So the GCD of the three, $x^5(x+y)^2(x+z)^8$, $16y^4z^7$, and **81** is the greatest of all the divisors common to all the three.

And getting the GCD, we get the product of all powers, in each of which the base is a prime factor common to all the three above, and the exponent is the smallest of all used for the prime factor.

• So we want to begin with factorizing **81** and $\mathbf{16y^4z^7}$. Then, we get: $\mathbf{3^4}$ and $\mathbf{2^4y^4z^7}$.

Thus, we are getting the GCD of $\mathbf{3^4}$, $\mathbf{2^4y^4z^7}$, and $x^5(x+y)^2(x+z)^8$.

• Next, finding all prime factors common to all the three above, we get none this time.

Thus in this case, we get no power where the base is a prime factor common to all the three above, so we don't need to take such a product as mentioned above.

What is the GCD, though?

There is a divisor that can divide any value no matter what the value may be, and the divisor is 1, and no other can do so. Thus in this case, 1 is the GCD.

Let's next, move on to the LCM, which is a multiple common and the smallest.

A multiple common to a set of objects as integers has <u>at least</u> all the prime factors all the objects have, and the exponent to be used for each prime factor is <u>at least</u> the largest of all used for the prime factor.

So the LCM, the multiple the <u>least</u> and common to all the objects has *all prime factors* all the objects have, and the *exponent* to be used for each prime factor is *the largest* of all used for the prime factor.

Now, in the case of 3^4, $2^4y^4z^7$, and $x^5(x+y)^2(x+z)^8$, all prime factors used are: $\mathbf{2, 3, x, y,}$ $\mathbf{z, x+y,}$ and $\mathbf{x+z,}$

4 is the largest exponent used for each of 2 and 3,

5 is the largest exponent used for x,

4 is the largest exponent used for y,

7 is the largest exponent used for z,

2 is the largest exponent used for $x + y$,

and 8 is the largest exponent used for $x + z$.

So a multiple common to the three above has as prime factors <u>at least</u> $\mathbf{2, 3, x, y, z, x+y,}$ and $x + z$, and the exponents to be used for the prime factors are <u>at least</u> 4, 4, 5, 4, 7, 2, and 8 respectively.

Therefore, the LCM is: $\mathbf{2^4 3^4 x^5 y^4 z^7 (x+y)^2 (x+z)^8}$.

Suggestions or Solutions
To the **Problem** in the Example **2**

Find the GCD and LCM of x, y, and z.

GCD is a divisor common and the largest. So the GCD of the three monomials given in the problem is the greatest of all the divisors common to all the three.

To begin with, we want to find all the prime factors common to all the three monomials. This time however, all the three monomials themselves are all prime.
What then, is the GCD?

Every object except 0 and 1 has at least two divisors, which are 1 and itself.
In particular, 1 is a divisor common to every object.
So what are divisors common to x, y, and z?

It is 1 only, and therefore, the GCD of x, y, and z is 1.

Let's next, move on to the LCM, which is a multiple common and the smallest.

A common multiple of a set of monomials has <u>at least</u> all the prime factors all the monomials have, and the exponent to be used for each prime factor is <u>at least</u> the largest of all used for the prime factor.

So the LCM, the multiple the <u>least</u> and common to a set of monomials has *all prime factors* all the monomials have, and the *exponent* to be used for each prime factor is *the largest* of all used for the prime factor.

Now, in this case, the three monomials x, y, and z themselves are all prime, 1 is the largest exponent used for each of x, y, and z.

Therefore, a multiple common to x, y, and z has as prime factors <u>at least</u> x, y, and z, and the exponent to be used for each of the prime factors is <u>at least</u> 1.

Thus, the LCM of x, y, and z is: xyz.

Suggestions or Solutions
To the **Problem** in the Example **3**

Find the GCD and LCM of $x^2 + 3x + 2$ and $x^2 - 1$.

GCD is a divisor common and the largest. So the GCD of a set of polynomials is the greatest of all the divisors common to all the polynomials.

And getting the GCD, we get the product of all powers, in each of which the base is a prime factor common to all the polynomials, and the exponent is the smallest of all used for the prime factor.

- So we want to begin with factorizing the two polynomials $x^2 + 3x + 2$ and $x^2 - 1$.

Then, we get: $x^2 + 3x + 2 = (x + 1)(x + 2)$, and $x^2 - 1 = (x + 1)(x - 1)$.

- Next, finding all the prime factors common to both polynomials, we get: $x + 1$.

- Next, finding the smallest of all exponents used for the prime factor above, we get 1.

Thus, in this case, we get only one power where the base is a prime factor common to both polynomials, so we don't need to take such a product as stated above.

Thus, the GCD is just a power, where the base is: $x + 1$, which is the prime factor, and the exponent is the smallest of all exponents used for the prime factor $x + 1$.

The smallest exponent is 1. So the GCD is: $x + 1$.

Let's next, move on to the LCM, which is a multiple common and the smallest.

A common multiple of a set of polynomials has <u>at least</u> all the prime factors all the polynomials have, and the exponent to be used for each prime factor is <u>at least</u> the largest of all used for the prime factor.

So the LCM, the multiple the <u>least</u> and common to all the polynomials has *all prime factors* all the polynomials have, and the *exponent* to be used for each prime factor is *the largest* of all used for the prime factor.

Now, in the case of two polynomials $x^2 + 3x + 2$ and $x^2 - 1$, factorized to $(x + 1)(x + 2)$ and $(x + 1)(x - 1)$, all prime factors used are: $x + 1$, $x + 2$, and $x - 1$.

And 1 is the largest exponent used for each of all the prime factors.

So a multiple common to both polynomials has as prime factors <u>at least</u> $x + 1$, $x + 2$, and $x - 1$, and the exponent to be used for each of the prime factors is <u>at least</u> 1.

Thus, the LCM is: $(x + 1)(x + 2)(x - 1)$.

Suggestions or Solutions
To the **Problem** in the Example **4**

Find the GCD and LCM of polynomials below:
$(x + 1)^2(x + 2)(x + 3)^2$, $\quad x(x - 1)(x + 1)^4(x + 3)^3$, $\quad$ **and** $(x + 2)^3(x + 1)x^3(x + 3)^3$.

The GCD of a set of polynomials is the greatest divisor common to all the polynomials.

And getting the GCD, we get the product of all powers, in each of which the base is a prime factor common to all the polynomials, and the exponent is the smallest of all used for the prime factor.

Now, we want to find the GCD of three polynomials, which are fully factorized already. So we are ready to get the GCD.

• First, finding all the prime factors common to the three polynomials, we get:

$x + 1$ and $x + 3$.

• Next, we take the product of all the prime factors above, and apply the smallest of all the exponents used for each of the prime factors.

Then, 1 is the smallest of all the exponents used for the prime factor $x + 1$, and 2 is the smallest used for the prime factor $x + 3$, so we get: $(x + 1)(x + 3)^2$, which is the GCD.

Let's next, move on to the LCM, which is a multiple common and the smallest.

A multiple common to a set of polynomials has <u>at least</u> all the prime factors all the polynomials have, and the exponent to be used for each prime factor is <u>at least</u> the largest of all used for the prime factor.

Therefore, the LCM, the <u>least</u> multiple common to a set of polynomials has *all prime factors* all the polynomials have, and the *exponent* to be used for each prime factor is *the largest* of all used for the prime factor.

Now, in the case of three polynomials $(x + 1)^2(x + 2)(x + 3)^2, \quad x(x - 1)(x + 1)^4(x + 3)^3,$ and $(x + 2)^3(x + 1)x^3(x + 3)^3$, all prime factors all the three polynomials have are: $x + 1$, $x + 2$, $x + 3$, x, and $x - 1$.

4 is the largest exponent used for $x + 1$,

3 is the largest exponent used for each of $x + 2$ and $x + 3$,

and 1 is the largest exponent used for each of x and $x - 1$.

So a multiple common to the three polynomials has as prime factors <u>at least</u> $x + 1$, $x + 2$, $x + 3$, x, and $x - 1$, and the exponents to be used for the prime factors are <u>at least</u> 4, 3, 3, 1, and 1 respectively.

Therefore, the LCM of the three polynomials is: $x(x + 1)^4(x + 2)^3(x + 3)^3(x - 1)$.

Examples 3 in GCD and LCM

Find the GCD (Greatest Common Divisor) and LCM (Least Common Multiple) in each of the cases below.

0. $(x + 1)^4(y - 1)$ and $(x + 2)(y + 1)^2$

1. $2x^4 + 20x^3 + 62x^2 + 60x$ and $x^4 + x^3 - 2x^2$

2. $t^2(t^2 - 9)(t + 5)^{-2}(t + 3)^2(t - 2)$ and $t^3(t + 3)^4(t - 3)^2(t - 1)(t + 5)^2$

3. $(9x - 3x^2 - 27)(x + 2)^3(x - 1)^3(x + 1)(x - 3)$ and $9(x^3 + 27)^2(x^2 + 4x + 4)(x^2 - 1)^2$

4. $12(x^2 - 1)^2(x^3 - 1)(x + 3)$ and $54(x + 1)^3(x + 3)^2(x^2 - x - 2)^2(x - 3)$

Suggestions or Solutions
To the **Problem** in the Example **0**

`

Find the GCD and LCM of $(x + 1)^4(y - 1)$ and $(x + 2)(y + 1)^2$.

Getting the GCD of a set of polynomials, we get the product of all powers, in each of which the base is a prime factor common to all the polynomials, and the exponent is the smallest of all used for the prime factor.

Now, we are given two polynomials, fully factorized already, so we are ready to get the GCD. Finding however, all the prime factors common to both polynomials, we get none. What then, is the GCD?

Every polynomial has at least two divisors, which are 1 and itself.
So what are divisors common to $(x + 1)^4(y - 1)$ and $(x + 2)(y + 1)^2$?

It is 1, which is the only divisor, and therefore, is the GCD. So 1 is the only divisor common to every polynomial. And the same is true, too, for integers, monomials, etc.

Let's next, move on to the LCM, which is a multiple common and the smallest.

A multiple common to a set of polynomials has <u>at least</u> all the prime factors all the polynomials have, and the exponent to be used for each prime factor is <u>at least</u> the largest of all used for the prime factor.

Thus, the <u>least</u> multiple common to all the polynomials has *all prime factors* all the polynomials have, and the *exponent* to be used for each prime factor is *the largest* of all used for the prime factor.

Now, in the case of two polynomials $(x + 1)^4(y - 1)$ and $(x + 2)(y + 1)^2$, all the prime factors the two have are: $x + 1$, $y - 1$, $x + 2$, x, and $y + 1$,

4 is the largest exponent used for $x + 1$,

1 is the largest exponent used for each of $y - 1$ and $x + 2$,

and 2 is largest exponent used for $y + 1$.

So a multiple common to the two polynomials has to have as prime factors <u>at least</u> $x + 1$, $y - 1$, $x + 2$, and $y + 1$, and the exponents to be used for the prime factors are <u>at least</u> 4, 1, 1, and 2 respectively in the order of $x + 1$, $y - 1$, $x + 2$, and $y + 1$.

Therefore, the LCM of the three polynomials is: $(x + 1)^4(y - 1)(x + 2)(y + 1)^2$.

Suggestions or Solutions

Find the GCD and LCM of $2x^4 + 20x^3 + 62x^2 + 60x$ and $x^4 + x^3 - 2x^2$.

The GCD of a set of polynomials is the product of all powers, in each of which the base is a prime factor common to all the polynomials, and the exponent is the smallest of all used for the prime factor. And the same is true for integers and monomials, too.

Now, in this problem, we are given two polynomials, which are not factorized yet.

• So first, we want to factorize the two polynomials.

To begin with, we get: $2x^4 + 20x^3 + 62x^2 + 60x = 2x(x^3 + 10x^2 + 31x + 30)$.

Next, if $x - d$ is a factor of $x^3 + 10x^2 + 31x + 30$, we can set:

$x^3 + 10x^2 + 31x + 30 = (x - d)Q$, where Q is a polynomials of degree 2 as $ax^2 + bx + c$.

So we get: $d^3 + 10d^2 + 31d + 30 = 0$.   How come?

That's because $x^3 + 10x^2 + 31x + 30 = (x - d)Q$, and $(d - d)Q = 0$.

How do we get such d, though?

Expanding (or simplifying) $(x - d)(ax^2 + bx + c)$, we get: $-dc = 30$.   How come?

Setting $Q = ax^2 + bx + c$, we get: $x^3 + 10x^2 + 31x + 30 = (x - d)(ax^2 + bx + c)$

$= ax^3 + bx^2 + cx - adx^2 - bdx - cd = ax^3 + (b - ad)x^2 + (c - bd)x - cd$.

Thus, we get: $a = 1$, $b - ad = 10$, $c - bd = 31$, and $-dc = 30$.

So d is a divisor of 30, and thus, is one of ± 1, ± 2, ± 3, ± 5, ± 6, ± 10, ± 15, and ± 30.

Now, we have set: $x^3 + 10x^2 + 31x + 30 = (x - d)(ax^2 + bx + c)$.

So putting a divisor of 30 into x in $x^3 + 10x^2 + 31x + 30$, and getting 0, the divisor is the value of d. And if any, we can find d by trial and error with divisors of 30.

What if no divisor can make it 0?

Then, d is not an integer, and $x - d$ where d is an integer is not a factor of the polynomial $x^3 + 10x^2 + 31x + 30$, so we can conclude that the polynomial has no factor, and is prime.

That is, $x^3 + 10x^2 + 31x + 30$ is a prime factor of $2x^4 + 20x^3 + 62x^2 + 60x$. So we can say that $2x^4 + 20x^3 + 62x^2 + 60x$ is fully factorized to $2x(x^3 + 10x^2 + 31x + 30)$.

Now, evaluating $x^3 + 10x^2 + 31x + 30$ for $x = -2$, we get:

$(-2)^3 + 10(-2)^2 + 31(-2) + 30 = -8 + 40 - 62 + 30 = -70 + 70 = 0$.

So $d = -2$, and thus, $x - d = x + 2$ is a divisor, that is, a factor of $x^3 + 10x^2 + 31x + 30$.

And next, we can get Q, which is the quotient we get dividing the polynomial by $x + 2$.

And getting the quotient, we can apply to the polynomial the synthetic division by $x + 2$.

Then, we get:

$$
\begin{array}{c|cccc}
-2 & 1 & 10 & 31 & 30 \\
   &   & -2 & -16 & -30 \\
\hline
   & 1 & 8 & 15 & 0
\end{array}
$$

Thus, dividing $x^3 + 10x^2 + 31x + 30$ by $x + 2$, we get: $x^2 + 8x + 15$ as the quotient.

So we get: $x^3 + 10x^2 + 31x + 30 = (x + 2)(x^2 + 8x + 15)$.

Next, we want to check to see if the quotient $x^2 + 8x + 15$ can still be factorized.

In fact, $x^2 + 8x + 15$ can be factorized to $(x + 5)(x + 3)$.

That is, $2x^4 + 20x^3 + 62x^2 + 60x = 2x(x^3 + 10x^2 + 31x + 30) = 2x(x + 2)(x + 5)(x + 3)$.

Therefore, $2x^4 + 20x^3 + 62x^2 + 60x$ gets factorized to $2x(x + 2)(x + 5)(x + 3)$.

Next, moving on to $x^4 + x^3 - 2x^2$, we get: $x^4 + x^3 - 2x^2 = x^2(x^2 + x - 2) = x^2(x + 2)(x - 1)$.

• So factorizing the two polynomials given in the problem, we get:
$2x(x + 2)(x + 5)(x + 3)$, and $x^2(x + 2)(x - 1)$.

• Next, finding all the prime factors common to both polynomials, we get: x and $x + 2$.

• Next, we find the smallest of all exponents used for each of the prime factors above. Then, 1 is the smallest exponent used for each of the prime factors x and $x + 1$.

• Next, we take the product of all the prime factors above, and apply the smallest of all exponents used for each of the prime factors.

Then, we get: $x(x + 1)$, which is the GCD of $2x(x + 2)(x + 5)(x + 3)$, and $x^2(x + 2)(x - 1)$.

••• Let's next, move on to the LCM, which is the multiple common and the smallest.

A multiple common to a set of polynomials has <u>at least</u> all prime factors all the polynomials have, and the exponent to be used for each prime factor is <u>at least</u> the largest of all used for the prime factor.

So the multiple the <u>least</u> and common to a set of polynomials has *all prime factors* all the polynomials have, and the *exponent* to be used for each prime factor is *the largest* of all used for the prime factor.
And the same is true for integers and monomials, too.

• So first, finding <u>all the prime factors</u> both polynomials have, we get:

$x, x + 2, x + 5, x + 3$, and $x - 1$.

• Next, we take the product of all the prime factors above since the LCM has all the prime factors both polynomials have.

Then, we can simply get: $x(x + 2)(x + 5)(x + 3)(x - 1)$.

• Next, we find <u>the largest exponent</u> of all used for each of the prime factors above.

Then, 2 is the largest used for x, and 1 is the one used for each of $x + 2, x + 5, x + 3$, and $x - 1$.

Therefore, the LCM is: $x^2(x + 2)(x + 5)(x + 3)(x - 1)$.

Suggestions or Solutions
To the **Problem** in the Example **2**

Find the GCD and LCM of two polynomials below:
$t^2(t^2 - 9)(t + 5)^{-2}(t + 3)^2(t - 2)$ **and** $t^3(t + 3)^4(t - 3)^2(t - 1)(t + 5)^2$.

The two polynomials given look already factorized. The first of the two is however, not fully factorized. It has: $t^2 - 9$, which can get factorized to $(t + 3)(t - 3)$. So we get:

$$t^2(t^2 - 9)(t + 5)^{-2}(t + 3)^2(t - 2) = t^2(t + 3)(t - 3)(t + 5)^{-2}(t + 3)^2(t - 2)$$
$$= t^2(t + 3)^3(t - 3)(t + 5)^{-2}(t - 2).$$

Thus, we now have: $t^2(t + 3)^3(t - 3)(t + 5)^{-2}(t - 2)$ and $t^3(t + 3)^4(t - 3)^2(t - 1)(t + 5)^2$.

• So first, finding the GCD, we want to find first, all the prime factors common to both polynomials. Then, we get: $t, t + 3, t - 3$, and $t + 5$.

• Next, we find the smallest of all exponents used for each of the prime factors above.

Then, 2 is the smallest exponent used for each of the prime factors t, 3 is the smallest used for $t + 3$, 1 is the one for $t - 3$, and -2 is used for $t + 5$.

• Next, we take the product of all the prime factors above, and then, apply to each of the factors the smallest of all exponents used for the factor.

Then, we get: $t^2(t + 3)^3(t - 3)(t + 5)^{-2}$, which is the GCD.

Let's next, move on to the LCM, which is a multiple common and the smallest.

• First, we take the product of <u>all the prime factors</u> all the polynomials have.

Then, we can simply get: $t(t + 3)(t - 3)(t + 5)(t - 2)(t - 1)$.

• Next, we apply to each factor above <u>the largest exponent</u> of all used for the factor.

Then, 3 is the largest for the factor t, 4 is used for the factor $t + 3$, 2 is used for each of the factors $t - 3$ and $t - 5$, and 1 is used for each of $t - 2$ and $t - 1$.

Therefore, the LCM is: $t^3(t + 3)^4(t - 3)^2(t + 5)^2(t - 2)(t - 1)$.

Suggestions or Solutions
To the **Problem** in the Example **3**

Find the GCD and LCM of two polynomials below:
$$(9x - 3x^2 - 27)(x + 2)^3(x - 1)^3(x + 1)(x - 3) \text{ and } 9(x^3 + 27)^2(x^2 + 4x + 4)(x^2 - 1)^2.$$

- To begin with, we want to factorize the two polynomials given.

Though they both look already factorized, they are not fully factorized.

The first one has: $9x - 3x^2 - 27$, which can still be factorized.
So factorizing it, we get: $9x - 3x^2 - 27 = -3(x^2 - 3x + 9)$.

Next, the second one has: $x^2 + 4x + 4$ and $x^2 - 1$, both of which can still be factorized.

Factorizing them both, we get: $x^2 + 4x + 4 = (x + 2)^2$, and $x^2 - 1 = (x + 1)(x - 1)$.

And also, $9(x^3 + 27)^2(x^2 + 4x + 4)(x^2 - 1)^2$ has: $x^3 + 27$, which can still be factorized, too. How then, can we get it factorized?

We have a factorization identity: $a^3 + b^3 = (a + b)(a^2 - ab + b^2)$.

So using the identity, we can get: $x^3 + 27 = x^3 + 3^3 = (x + 3)(x^2 - 3x + 9)$. Thus, we get:

$$9(x^3 + 27)^2(x^2 + 4x + 4)(x^2 - 1)^2 = 9(x + 3)^2(x^2 - 3x + 9)^2(x + 2)^2(x + 1)^2(x - 1)^2.$$
In addition, we have $9 = 3^2$, too.

Therefore, factorizing fully the two polynomials given, we get:

$-3(x^2 - 3x + 9)(x + 2)^3(x - 1)^3(x + 1)(x - 3)$ and
$3^2(x + 3)^2(x^2 - 3x + 9)^2(x + 2)^2(x + 1)^2(x - 1)^2$.

- So next, finding all the prime factors common to the two polynomials, we get:
$3, x^2 - 3x + 9, x + 2, x - 1$, and $x + 1$.

• Next, we find the smallest of all exponents used for each of the prime factors above.

Then, 1 is the smallest exponent used for each of 3 and $x^2 - 3x + 9$, 2 is the one for each of $x + 2$ and $x - 1$, and again, 1 is the one used for $x + 1$.

• Next, we take the product of all the prime factors above, and apply the smallest of all exponents used for each of the prime factors.

Then, we get: $3(x^2 - 3x + 9)(x + 2)^2(x - 1)^2(x + 1)$, which is the GCD.

Let's next, move on to the LCM, which is a multiple common and the smallest.

• First, we want to collect <u>all the prime factors</u> all the polynomials have, and take the product of the factors.

Then, we can simply get: $3(x^2 - 3x + 9)(x + 2)(x - 1)(x + 1)(x - 3)(x + 3)$.

• Next, we want to get <u>the largest exponents</u> of all used for all the prime factors above.

Then, 2 is the largest used for each of 3, $x + 1$, and $x^2 - 3x + 9$, 3 is the one used for each of $x + 2$ and $x - 1$, 1 is the one used for $x - 3$, and 2 is the one used for $x + 3$.

Therefore, the LCM is: $3^2(x^2 - 3x + 9)^2(x + 2)^3(x - 1)^3(x + 1)^2(x - 3)(x + 3)^2$.

We can put it this way, too: $9(x^2 - 3x + 9)^2(x + 2)^3(x^2 - 1)^2(x - 1)(x^2 - 9)(x + 3)$.
That's because:
$(x - 1)^3(x + 1)^2 = (x - 1)(x - 1)^2(x + 1)^2 = (x - 1)\{(x - 1)(x + 1)\}^2 = (x - 1)(x^2 - 1)^2$.
$(x - 3)(x + 3)^2 = (x - 3)(x + 3)(x + 3) = (x^2 - 9)(x + 3)$.

We can put it this way, too: $9(x^2 - 3x + 9)^2(x + 2)^3(x^2 - 1)^2(x^2 - 9)(x^2 + 2x - 3)$.
That's because $(x - 1)(x + 3) = x^2 + 2x - 3$.

We can put it this way, too: $9(x^2 - 3x + 9)^2(x + 2)^3(x^4 - 10x^2 + 9)(x^2 - 1)(x^2 + 2x - 3)$.

That's because $(x^2 - 1)^2(x^2 - 9) = (x^2 - 1)(x^2 - 1)(x^2 - 9) = (x^2 - 1)(x^4 - 10x^2 + 9)$.

Suggestions or Solutions
To the **Problem** in the Example **4**

Find the GCD and LCM of two polynomials below:
$$12(x^2 - 1)^2(x^3 - 1)(x + 3) \text{ and } 54(x + 1)^3(x + 3)^2(x^2 - x - 2)^2(x - 3).$$

• First, we want to factorize the two polynomials given.

Though they both look already factorized, they are not fully factorized.

The first one has $x^2 - 1$ and $x^3 - 1$, both of which can still be factorized. Factorizing them both, we get:

$$x^2 - 1 = (x + 1)(x - 1), \text{ and } x^3 - 1 = (x - 1)(x^2 + x + 1).$$

Besides, we have: $12 = 2^2 3$, too.

Thus, we get: $12(x^2 - 1)^2(x^3 - 1)(x + 3) = 2^2 3(x + 1)(x - 1)(x - 1)(x^2 + x + 1)(x + 3)$

$$= 2^2 3(x + 1)(x - 1)^2(x^2 + x + 1)(x + 3).$$

Next, the second one has $x^2 - x - 2$, which can still be factorized.

Factorizing it, we get: $x^2 - x - 2 = (x - 2)(x + 1)$.

In addition, we have: $54 = 2 \cdot 3^3$, too.

Thus, $54(x + 1)^3(x + 3)^2(x^2 - x - 2)^2(x - 3) = 2 \cdot 3^3(x + 1)^3(x + 3)^2(x - 2)(x + 1)(x - 3)$

$$= 2 \cdot 3^3(x + 1)^4(x + 3)^2(x - 2)(x - 3).$$

And thus, fully factorizing the two polynomials given, we get:

$$2^2 3(x + 1)(x - 1)^2(x^2 + x + 1)(x + 3) \text{ and } 2 \cdot 3^3(x + 1)^4(x + 3)^2(x - 2)(x - 3).$$

So let's now begin with the GCD of the two above.

• First, finding all the prime factors common to the two polynomials, we get:

$2, 3, x + 1$, and $x + 3$.

• Next, we take the product of all the prime factors above, and apply to each prime factor the smallest of all exponents used for the prime factor.

Then, we get: $2^1 \cdot 3^1 (x + 1)^1 (x + 3)^1$, which equals $6(x + 1)(x + 3)$, which is the GCD.

Let's next, move on to the LCM, which is a multiple common and the smallest.

• First, we want to collect <u>all the prime factors</u> all the polynomials have.

Then, we get: $2, 3, x + 1, x - 1, x^2 + x + 1, x + 3, x - 2$, and $x - 3$.

• Next, we get <u>the largest exponents</u> of all used for all the prime factors above.

Then, 2 is the largest used for each of $2, x - 1$, and $x + 3$, 3 is the one used for 3, 4 is the one used for $x + 1$, and 1 is the one used for each of $x^2 + x + 1, x - 2$, and $x - 3$.

• Next, we want to take the product of all the prime factors above, and apply the exponents above to those prime factors respectively.

Then, we get: $2^2 3^3 (x^2 + x + 1)(x - 1)^2 (x + 3)^2 (x + 1)^4 (x - 2)(x - 3)$, which is the LCM.

Examples 4 in GCD and LCM

0. Find the GCD and LCM of 300 and 24255

1. Find the GCD and the LCM of the polynomials as follows:

$ab^2xy^2(x + y)(x + 2y)(x + 3y),$ $a^2bx^2y(x + y)(x + 2y),$ $abx^3y^2(x + 1)(x + y)^2(x + 3y),$

$xy(x + y)^2,$ and $ax^2y^3(x + y)(x + 2y)^2.$

2. Assuming that G is the GCD of A and B, and that L is the LCM of A and B, show that $LG = AB$.

3. Suppose dividing A by B, we get Q as the quotient and R as the remainder.

Then, we get: $A = BQ + R$. Show that the GCD of A and B = the GCD of B and R.

Suggestions or Solutions
To the **Problem** in the Example **0**

Find the GCD and LCM of 300 and 24255.

Finding the GCD of a set of integers, we usually take a product. What product is it?

It is a product of all powers, in each of which the base is a prime factor common to all the integers, and the exponent is the smallest of all used for the prime factor.

So the product has all the prime factors common to all the integers, and thus, is the divisor the greatest and common to all the integers, that is, the GCD.

So in short, finding the GCD of a set of integers, we get all the prime factors common to all the integers, together with the exponents the smallest of all used for the prime factors, and then, take the product of the powers.

• Thus, we want to begin with factorizing 300 and 24255.   Then, we get:

$300 = 3 \cdot 100 = 3 \cdot 4 \cdot 25 = 2^2 3^1 5^2 = 2^2 3 \cdot 5^2$.

$24255 = 24000 + 240 + 15 = 3 \cdot 8000 + 3 \cdot 80 + 3 \cdot 5 = 3 \cdot (8 \cdot 1000 + 8 \cdot 10 + 5)$

$= 3(8 \cdot 5 \cdot 200 + 8 \cdot 2 \cdot 5 + 5) = 3 \cdot 5(1600 + 16 + 1) = 3 \cdot 5(1500 + 90 + 27)$

$= 3 \cdot 5(3 \cdot 500 + 3^2 10 + 3^3) = 3^2 5(500 + 30 + 9) = 3^2 5(490 + 49) = 3^2 5 \cdot 7(70 + 7)$

$= 3^2 5 \cdot 7^2(10 + 1) = 3^2 5^1 7^2 11^1 = 3^2 5 \cdot 7^2 11$.

• Next, finding all prime factors common to all the integers, we get 3 and 5.

• Next, we find the smallest of all exponents used for each of the prime factors above.

Then, 1 is the smallest exponent used for each of 3 and 5.

So we get two powers, in one of which, the base is 3, and the exponent is 1, and in the other, the base is 5, and the exponent is 1, too. That is, we get 3^1 and 5^1.

Thus, taking the product of two powers, we get the GCD, which is $3^1 5^1$, which is 15.

Next, what is the LCM of a set of integers?

To begin with, a multiple common to a set of integers has <u>at least</u> all prime factors all the integers have, and the exponent to be used for each of the prime factors is <u>at least</u> the largest of all used for the prime factor.

So the LCM, <u>the least</u> of all multiples common to all the integers has <u>all prime factors</u> all the integers have, and the <u>exponent</u> to be used for each of the prime factors is <u>the largest</u> of all used for the prime factor.

• Thus, finding the LCM of a set of integers, too, we want to begin with factorizing the integers. We have already done so, and they are:

$300 = 2^2 3^1 5^2$, and $24255 = 3^2 5 \cdot 7^2 11$.

• Next, finding all the prime factors both integers have, we get: **2**, **3**, **5**, **7**, and **11**.

• Next, find the largest exponent used for each of the prime factors above.

Then, 2 is the largest used for each of **2**, **3**, **5**, and **7**, and 1 is the one used for **11**.

Therefore, a multiple common to both integers has as prime factors <u>at least</u> **2**, **3**, **5**, **7**, and **11**, and the exponents to be used for the prime factors are <u>at least</u> 2, 2, 2, 2, and 1 respectively.

Thus, the LCM, that is, the least common multiple is: $2^2 3^2 5^2 7^2 11$, which is 485100.

Suggestions or Solutions

Find the GCD and the LCM of the polynomials as follows:

$ab^2xy^2(x+y)(x+2y)(x+3y)$, $\quad a^2bx^2y(x+y)(x+2y)$, $\quad abx^3y^2(x+1)(x+y)^2(x+3y)$, $xy(x+y)^2$, $\quad$ **and** $ax^2y^3(x+y)(x+2y)^2$.

Finding the GCD of a set of polynomials, we usually take a product. What product is it?

It is a product of all powers, in each of which the base is a prime factor common to all the polynomials, and the exponent is the smallest of all used for the prime factor.

So the product has all the prime factors common to all the polynomials, and thus, is the divisor the greatest and common to all the polynomials, that is, the GCD.

Therefore in short, finding the GCD of a set of polynomials, we get all the prime factors common to all the polynomials, together with the exponents the smallest of all used for the prime factors, and then, take the product of the powers.

• Thus, we want to begin with factorizing the polynomials given.
They all have already been fully factorized.

• So next, we want to find all the prime factors common to all the polynomials.
Then, we get: x, y, and $x+y$.

• Next, find the smallest of all the exponents used for each of the prime factors above.
Then, 1 is the smallest exponent used for each of all the prime factors above.

So we get three powers, in one of which, the base is x, and the exponent is 1, in another, the base is y, and the exponent is 1, and in the other, the base is $x+y$, and the exponent is 1, too.
That is, we get: x^1, y^1, and $(x+y)^1$, which are just x, y, and $x+y$, of course.
Thus, taking the product of the three powers, we get the GCD, which is: $x^1y^1(x+y)^1$, which is: $xy(x+y)$.

Next, what is the LCM of a set of polynomials?

To begin with, a multiple common to a set of polynomials has <u>at least</u> all prime factors all the polynomials have, and the exponent to be used for each of the prime factors is <u>at least</u> the largest of all used for the prime factor.

So the LCM, <u>the least</u> of all multiples common to all the polynomials has <u>all prime factors</u> all the polynomials have, and the <u>exponent</u> to be used for each of the prime factors is <u>the largest</u> of all used for the prime factor.

• Thus, finding the LCM of a set of polynomials, too, we want to begin with factorizing the polynomials.

They all have already been fully factorized, and they are:

$ab^2xy^2(x+y)(x+2y)(x+3y)$, $a^2bx^2y(x+y)(x+2y)$, $abx^3y^2(x+1)(x+y)^2(x+3y)$, $xy(x+y)^2$,   and $ax^2y^3(x+y)(x+2y)^2$.

• So next, finding all the prime factors all the polynomials have, we get:

a, b, x, y, $x+1$, $x+y$, $x+2y$,   and $x+3y$.

• Next, we want to find the largest exponent used for each of the prime factors above.

Then, 2 is the largest used for each of a, b, $x+y$, and $x+2y$, and 1 is the one used for each of $x+1$ and $x+3y$, and 3 is the one used for each of x and y.

Therefore, a multiple common to all the polynomials has as prime factors <u>at least</u> a, b, x, y, $x+1$, $x+y$, $x+2y$, and $x+3y$, and the exponents to be used for the prime factors are <u>at least</u> 2, 2, 3, 3, 1, 2, 2, and 1 respectively.

Thus, the LCM is: $a^2b^2x^3y^3(x+1)(x+y)^2(x+2y)^2(x+3y)$.

Suggestions or Solutions
To the **Problem** in the Example **2**

Assuming that *G* is the GCD of *A* and *B*, and that *L* is the LCM of *A* and *B*, show that $LG = AB$.

Suppose *a* and *b* are prime to each other.

Then, $A = aG$, and $B = bG$, since *G* is the GCD of *A* and *B*.

So $L = abG$, and therefore, $LG = abGG = aGbG = AB$.

If not quite sure of the idea behind the processes above, follow the steps below:

To begin with, what is GCD?

It is the divisor common and the greatest. So for instance, the GCD of a set of integers is the greatest of all divisors common to all the integers.

Suppose $A = aG$, $B = bG$, and *a* and *b* are prime to each other, that is, there is no factor common to *a* and *b*.   Then, *G* is the GCD of *A* and *B*.

Next, what is LCM?

It is the multiple common and the least. So for instance, the LCM of a set of integers is the least of all multiples common to all the integers in the set.

Suppose now, *L* is the LCM of *A* and *B*. Then, $L = abG$.   How come?

To begin with, we have a fact below:

• For instance, a multiple common to a set of integers is a product of powers, and has <u>at least</u> all the prime factors all the integers have, and the exponent to be used for each prime factor is <u>at least</u> the largest of all the exponents used for the prime factor.

So the LCM, the least of all multiples common to all the integers has all prime factors all the integers have, and the exponent to be used for each prime factor is the largest of all used for the prime factor.

And thus, getting the LCM of A and B, we get the product of all prime factors that A and B have, and then, apply to each factor the largest of all the exponents used for the factor.

Now, suppose first, a, b, and G are prime to each other.   For instance, $G = cd$.

Then, we get: $A = aG = acd$, and $B = bG = bcd$.

We know getting the LCM of A and B, we get the product of all prime factors that A and B have, and then, apply to each factor the largest of all the exponents used for the factor.

So the LCM of A and B is: $abcd$, since all the largest exponents are 1.

Thus, we get: $L = abG$, since $G = cd$.

Next, suppose a and G are not prime to each other, for instance, $G = ac$.

Then, we get: $A = aG = aac = a^2c$, and $B = bG = bac$.

So we can say that a, b, and c are all the prime factors that A and B have.

Next, the largest exponent used for a is 2, the one used for each of b and c is 1.

Thus, we get: $L = a^2bc$. And we know $G = ac$.

So we get: $L = a^2bc = abac = abG \Rightarrow L = abG$.

We will get the same result, too, in the case where b and G are not prime to each other, for instance, $G = bc$.

So putting threads together, we get: $L = abG$ if $A = aG$ and $B = bG$.

Thus, we get: $LG = abGG = aGbG = AB$.

In short:

Suppose a and b are prime to each other.

Then, $A = aG$, and $B = bG$, since G is the GCD of A and B.

So $L = abG$, and therefore, $LG = abGG = aGbG = AB$.

What if G is the GCD of A, B, and C, and L is the LCM of A, B, and C?

Then, we get: $LG^2 = ABC$.   How come?

Suppose a, b, and c are prime to each other.

Then, $A = aG$, $B = bG$, and $C = cG$, since G is the GCD of A, B, and C.

So $L = abcG$, and therefore, $LG^2 = abcGG^2 = aGbGcG = ABC$.

Thus by the same token:

Suppose G is the GCD of A_1, A_2, ... and A_n, and L is the LCM of A_1, A_2, ... and A_n.

Then, we get: $LG^{n-1} = A_1A_2 ... A_n$.   How come?

Suppose a_1, a_2, ... and a_n are prime to each other.

Then, $A_1 = a_1G$, $A_2 = a_2G$, ..., and $A_n = a_nG$, since G is the GCD of A_1, A_2, ..., and A_n.

Thus, $L = a_1a_2 ... a_nG$, so $LG^{n-1} = a_1a_2 ... a_nGG^{n-1} = a_1Ga_2 ... a_nG = A_1A_2 ... A_n$.

Suggestions or Solutions
To the **Problem** in the Example **3**

Suppose dividing A by B, we get Q as the quotient and R as the remainder. Then, $A = BQ + R$. Show that the GCD of A and B = the GCD of B and R.

Suppose $A = aG$, $B = bG$, and a and b are prime to each other.

Then, G is the GCD of A and B, and we get:

$$A = BQ + R \Rightarrow aG = bGQ + R \Rightarrow R = (a - bQ)G.$$

So we now have: $B = bG$, and $R = (a - bQ)G$.

Thus, G is a divisor common to B and R.

Therefore, if showing that b and $(a - bQ)$ are prime to each other, we show that G is not just a divisor common to B and R but the GCD of B and R, too.

Now, suppose that b and $(a - bQ)$ are **not** prime to each other.

In other words, we assume that G is **not** the GCD of B and R.

Then, there is a factor common to b and $(a - bQ)$.

Suppose now, the common factor is m.
Then, we can set: $b = mk$, and $a - bQ = mp$.

Thus, we get: $a - bQ = mp \Rightarrow a = bQ + mp = mkQ + mp = m(kQ + p)$, since $b = mk$.

So we now have: $a = m(kQ + p)$, and $b = mk$.

Thus, m is a common divisor of a and b, which however, contradicts the fact that a and b are prime to each other.

So it is not true that b and $(a - bQ)$ are **not** prime to each other.

That is, b and $(a - bQ)$ are prime to each other.

So G is the GCD of B and R, since $B = bG$ and $R = (a - bQ)G$.

And we know that G is the GCD of A and B.

Therefore, the GCD of A and B is the GCD of B and R.

Examples 5 in GCD and LCM

Suppose $A = x^4 + 2x^3 - x - 2$, $B = 2x^4 + 3x^3 - x^2 - 4$, and $C = x^3 + 7x^2 + 11x + 2$.

Suppose also, $A \& B = G$, where G is the GCD of A and B.   Then, find:

0. $(\frac{A}{A \& B}) \& (\frac{B}{A \& B})$, and $(A \& B) \& \{(\frac{A}{A \& B}) \& (\frac{B}{A \& B})\}$.

1. $(A \& B) \& A$, $(A \& C) \& B$, $(B \& C) \& A$, and $(B \& C) \& B$.

Suggestions or Solutions
To the **Problem** in the Example **0**

Suppose $A = x^4 + 2x^3 - x - 2$, $B = 2x^4 + 3x^3 - x^2 - 4$, and $C = x^3 + 7x^2 + 11x + 2$.

Suppose also, $A \& B = G$, where G is the GCD of A and B. Then, find:

$(\frac{A}{A\&B}) \& (\frac{B}{A\&B})$, and $(A \& B) \& \{(\frac{A}{A\&B}) \& (\frac{B}{A\&B})\}$.

Suppose $A = aG$, and $B = bG$.

Then, a and b are prime to each other, and we get: $\frac{A}{A\&B} = \frac{aG}{G} = a$, and $\frac{B}{A\&B} = \frac{bG}{G} = b$.

Thus, $(\frac{A}{A\&B}) \& (\frac{B}{A\&B}) = a \& b = 1$.

So next, we get: $(A \& B) \& \{(\frac{A}{A\&B}) \& (\frac{B}{A\&B})\} = (A \& B) \& 1 = G \& 1 = 1$.

If not quite sure of the idea behind the processes above, follow the steps below:

What in the world is &?

'&' in this problem is nothing but an operator, which takes two operands, and produces a result, which is the GCD of the operands.

Thus, in $(\frac{A}{A\&B}) \& (\frac{B}{A\&B})$, $\frac{A}{A\&B}$ and $\frac{B}{A\&B}$ are operands, so $(\frac{A}{A\&B}) \& (\frac{B}{A\&B})$ is no more than the GCD of $\frac{A}{A\&B}$ and $\frac{B}{A\&B}$.

Next, we have: $A \& B = G$, where G is the GCD of A and B.

So suppose now, $A = aG$, and $B = bG$.

Then, a and b are prime to each other, since G is the GCD of A and B, so we get:

$\frac{A}{A\&B} = \frac{aG}{G} = a$, and $\frac{B}{A\&B} = \frac{bG}{G} = b$, since $A \& B = G$, where G is the GCD of A and B.

Thus, we get: $(\frac{A}{A\&B}) \& (\frac{B}{A\&B}) = a \& b$, which is just 1 since a and b are prime to each other.

How come do we get 1 as the GCD?

First of all, 1 can divide any number, constant, monomial, polynomial, etc.

Next, since a and b are prime to each other, there is no factor common to both.

That is, other than 1, there is no divisor common to both, so the GCD of both is 1.

- Let's next, move on to $(A \& B) \& \{(\frac{A}{A\&B}) \& (\frac{B}{A\&B})\}$.

First, assuming that $A = aG$, and that $B = bG$, we get: $(\frac{A}{A\&B}) \& (\frac{B}{A\&B}) = a \& b = 1$.

So we get: $(A \& B) \& \{(\frac{A}{A\&B}) \& (\frac{B}{A\&B})\} = (A \& B) \& 1$.

Next, the GCD of 1 and any object nonzero as 2, x, or $2x + y$ is just 1 no matter what the nonzero object may be.

Only 1 can divide 1, and divisors of an object nonzero include 1, too.

So we can see that $(A \& B) \& \{(\frac{A}{A\&B}) \& (\frac{B}{A\&B})\} = 1$.

Suggestions or Solutions
To the **Problem** in the Example **1**

Suppose $A = x^4 + 2x^3 - x - 2$, $B = 2x^4 + 3x^3 - x^2 - 4$, and $C = x^3 + 7x^2 + 11x + 2$.

Suppose also, A & $B = G$, where G is the GCD of A and B. Then, find:

$(A$ & $B)$ & A, $(A$ & $C)$ & B, $(B$ & $C)$ & A, and $(B$ & $C)$ & B.

Let's find **$(A$ & $B)$ & A** first.

To begin with, suppose $A = aG$, $B = bG$, and a and b are prime to each other.

Then, we get: **$(A$ & $B)$ & $A = G$ & $A = G$** since $A = aG$, and A & $B = G$, so G & $aG = G$.

In fact, we get **$(A$ & $B)$ & $B = G$**, too, since $B = bG$, and A & $B = G$, so G & $bG = G$.

So this time, we want to actually find G, which is the GCD of A and B.

Then first, we want to factorize all the three polynomials, which are as follows:

$A = x^4 + 2x^3 - x - 2$, $B = 2x^4 + 3x^3 - x^2 - 4$, and $C = x^3 + 7x^2 + 11x + 2$.

So let's now get going with A first.

To begin with, suppose $x - e$ is a factor of A.   Then, we can set:

$x^4 + 2x^3 - x - 2 = (x - e)Q$, where Q is a polynomials of degree 3 as $ax^3 + bx^2 + cx + d$.

So we get: $e^4 + 2e^3 - e - 2 = 0$.   How come?

That's because $x^4 + 2x^3 - x - 2 = (x - e)Q$, and $(e - e)Q = 0$.

How do we get such e, though?

Expanding (simplifying) $(x - e)(ax^3 + bx^2 + cx + d)$, in the result, we can get: $-ed = -2$, that is, we get: $ed = 2$.   How come?

Setting: $Q = ax^3 + bx^2 + cx + d$, we get:

$$x^4 + 2x^3 - x - 2 = (x - e)(ax^3 + bx^2 + cx + d)$$
$$= ax^4 + bx^3 + cx^2 + dx - aex^3 - bex^2 - cex - ed$$
$$= ax^4 + (b - ae)x^3 + (c - be)x^2 + (d - ce)x - ed.$$

Thus, we get: $a = 1$, $b - ad = 2$, $c - be = 0$, $d - ce = -1$, and $-ed = -2$.

Suppose now, d and e are integers. Then, d and e are divisors of 2.
Thus, e can be one of all the divisors ±1 and ±2.

So putting one of the divisors into x in $x^4 + 2x^3 - x - 2$, and getting 0, the one is the value of e. We can keep trying with each of all the divisors until we get 0.

So we can find e by trial and error with those divisors. What if no divisor can make it 0?

Then, e is not an integer, so $x - e$ where e is an integer is not a factor of $x^4 + 2x^3 - x - 2$.

We cannot just conclude though, that $x^4 + 2x^3 - x - 2$ has no factor, and thus, is prime.

That is to say that $x^4 + 2x^3 - x - 2$ could still be factorized other way.

Now, evaluating $x^4 + 2x^3 - x - 2$ for $x = 1$, we get: $1^4 + 2 \cdot 1^3 - 1 - 2 = 1 + 2 - 1 - 2 = 0$.

So $e = 1$, and thus, $x - e = x - 1$ is a divisor, that is, a factor of $x^4 + 2x^3 - x - 2$.

So next, doing the synthetic division by $x - 1$, we get:

$$
\begin{array}{r|rrrrr}
1 & 1 & 2 & 0 & -1 & -2 \\
  &   & 1 & 3 & 3 & 2 \\
\hline
  & 1 & 3 & 3 & 2 & 0 \\
\end{array}
$$

Thus, dividing $x^4 + 2x^3 - x - 2$ by $x - 1$, we get: $x^3 + 3x^2 + 3x + 2$ as the quotient.

So we get: $x^4 + 2x^3 - x - 2 = (x - 1)(x^3 + 3x^2 + 3x + 2)$.   What then, is the next?

We want to check to see if the quotient $x^3 + 3x^2 + 3x + 2$ can still be factorized. Then, by the same token, we can set:

$x^3 + 3x^2 + 3x + 2 = (x - d)Q$, where Q is a polynomials of degree 2 as $ax^2 + bx + c$.

So we can use the fact that if $d^3 + 3d^2 + 3d + 2 = 0$, $x - d$ is a divisor of $x^3 + 3x^2 + 3x + 2$.

And in fact, we can get: $x^3 + 3x^2 + 3x + 2 = 0$ for $x = -2$, which is a divisor of the numeric term 2 in the polynomial $x^3 + 3x^2 + 3x + 2$.

And actually, evaluating $x^3 + 3x^2 + 3x + 2$ for $x = -2$, we get:

$(-2)^3 + 3 \cdot (-2)^2 + 3(-2) + 2 = -8 + 12 - 6 + 2 = 0$.

So $x + 2$ is a divisor, that is, a factor of $x^3 + 3x^2 + 3x + 2$.

Thus next, doing the synthetic division by $x + 2$, we get:

$$
\begin{array}{r|rrrr}
-2 & 1 & 3 & 3 & 2 \\
   &   & -2 & -2 & -2 \\
\hline
   & 1 & 1 & 1 & 0 \\
\end{array}
$$

Thus, dividing $x^3 + 3x^2 + 3x + 2$ by $x + 2$, we get: $x^2 + x + 1$ as the quotient.

So we get: $x^3 + 3x^2 + 3x + 2 = (x + 2)(x^2 + x + 1)$.

Next, we want to check to see if the quotient $x^2 + x + 1$ can still be factorized.

In fact, $x^2 + x + 1$ cannot be factorized further.
That is, $x^2 + x + 1$ has no factor, and thus, is prime.

So $x^4 + 2x^3 - x - 2 = (x - 1)(x^3 + 3x^2 + 3x + 2) = (x - 1)(x + 2)(x^2 + x + 1)$.

Therefore, $A = x^4 + 2x^3 - x - 2$ gets factorized to $(x - 1)(x + 2)(x^2 + x + 1)$.

- Let's next, factorize $B = 2x^4 + 3x^3 - x^2 - 4$.

Then, by the same method for the polynomial A, we can set:

$2x^4 + 3x^3 - x^2 - 4 = (x - e)Q$, where Q is a polynomials of degree 3 as $ax^3 + bx^2 + cx + d$.

So we can use the fact that if $2e^4 + 3e^3 - e^2 - 4 = 0$, $x - e$ is a divisor of B.

Putting a divisor of -4 into x in $2x^4 + 3x^3 - x^2 - 4$, and getting 0, we can say that the divisor is e.
In fact, evaluating $2x^4 + 3x^3 - x^2 - 4$ for $x = 1$, we get: $2 \cdot 1^4 + 3 \cdot 1^3 - 1 - 4 = 5 - 5 = 0$.

So $x - 1$ is a divisor, that is, a factor of $2x^4 + 3x^3 - x^2 - 4$.

So next, doing the synthetic division by $x - 1$, we get:

1	2	3	-1	0	-4
		2	5	4	4
	2	5	4	4	0

Thus, dividing $2x^4 + 3x^3 - x^2 - 4$ by $x - 1$, we get $2x^3 + 5x^2 + 4x + 4$ as the quotient.

So we get: $2x^4 + 3x^3 - x^2 - 4 = (x - 1)(2x^3 + 5x^2 + 4x + 4)$.

Next, we want to check to see if the quotient $2x^3 + 5x^2 + 4x + 4$ can still be factorized.
Then, by the same token, we can set:

$2x^3 + 5x^2 + 4x + 4 = (x - d)Q$, where Q is a polynomials of degree 2 as $ax^2 + bx + c$. So we can use the fact that if $2d^3 + 5d^2 + 4d + 4 = 0$, $x - d$ is a divisor of $2x^3 + 5x^2 + 4x + 4$.

Putting a divisor of 4 into x in $2x^3 + 5x^2 + 4x + 4$, and getting 0, we can say that the divisor is d.

In fact, evaluating $2x^3 + 5x^2 + 4x + 4$ for $x = -2$, we get:

$2(-2)^3 + 5 \cdot (-2)^2 + 4(-2) + 4 = -16 + 20 - 8 + 4 = 0$.

So $x + 2$ is a divisor, that is, a factor of $2x^3 + 5x^2 + 4x + 4$.

So next, doing the synthetic division $x + 2$, we get:

$$
\begin{array}{r|rrrr}
-2 & 2 & 5 & 4 & 4 \\
   &   & -4 & -2 & -4 \\
\hline
   & 2 & 1 & 2 & 0
\end{array}
$$

Thus, dividing $2x^3 + 5x^2 + 4x + 4$ by $x + 2$, we get: $2x^2 + x + 2$ as the quotient.

So we get: $2x^3 + 5x^2 + 4x + 4 = (x + 2)(2x^2 + x + 2)$.   What then, is the next?

We want to check to see if the quotient $2x^2 + x + 2$ can still be factorized.

In fact, $2x^2 + x + 2$ cannot be factorized further, so it has no factor, and thus, is prime.

So $2x^4 + 3x^3 - x^2 - 4 = (x - 1)(2x^3 + 5x^2 + 4x + 4) = (x - 1)(x + 2)(2x^2 + x + 2)$.

Therefore, $B = 2x^4 + 3x^3 - x^2 - 4$ gets factorized to $(x - 1)(x + 2)(2x^2 + x + 2)$.

• Let's next, factorize $C = x^3 + 7x^2 + 11x + 2$.

Then, by the same method for the polynomial A or B, we can set:

$x^3 + 7x^2 + 11x + 2 = (x - d)Q$, where Q is a polynomials of degree 2 as $ax^2 + bx + c$.

So we can use the fact that if $d^3 + 7d^2 + 11d + 2 = 0$, $x - d$ is a divisor of C.

Putting a divisor of 2 into x in $x^3 + 7x^2 + 11x + 2$, and getting 0, we can say that the divisor is d.

In fact, evaluating $x^3 + 7x^2 + 11x + 2$ for $x = -2$, we get:

$(-2)^3 + 7(-2)^2 + 11(-2) + 2 = -8 + 28 - 22 + 2 = 0$.

So $x + 2$ is a divisor, that is, a factor of $x^3 + 7x^2 + 11x + 2$.

So next, doing the synthetic division by $x + 2$, we get:

```
-2 |  1     7     11     2
   |       -2    -10    -2
   +------------------------
      1     5      1     0
```

Thus, dividing $x^3 + 7x^2 + 11x + 2$ by $x + 2$, we get: $x^2 + 5x + 1$ as the quotient.

So we get: $x^3 + 7x^2 + 11x + 2 = (x + 2)(x^2 + 5x + 1)$.

Next, we want to check to see if the quotient $x^2 + 5x + 1$ can still be factorized.

In fact, $x^2 + 5x + 1$ cannot be factorized further, so it has no factor, and thus, is prime.

Therefore, $C = x^3 + 7x^2 + 11x + 2$ gets factorized to $(x + 2)(x^2 + 5x + 1)$.

So we now have:

    $A = x^4 + 2x^3 - x - 2$ factorized to $(x - 1)(x + 2)(x^2 + x + 1)$.

    $B = 2x^4 + 3x^3 - x^2 - 4$ factorized to $(x - 1)(x + 2)(2x^2 + x + 2)$.

    $C = x^3 + 7x^2 + 11x + 2$ factorized to $(x + 2)(x^2 + 5x + 1)$.

Thus, the GCD of A and B is $(x - 1)(x + 2)$.

So $(A \& B) \& A = (x - 1)(x + 2)$. In fact, $(A \& B) \& B = (x - 1)(x + 2)$, too. How come?

We know A **&** $B = G$, and $B = bG$.

So $(A$ **&** $B)$ **&** $B = G$ **&** $bG = G = (x - 1)(x + 2)$.

    • Let's next, move on to $(A$ **&** $C)$ **&** B.

We know A **&** C is the GCD of A and C. So do we actually want to get that GCD, first?

Not necessarily.

Suppose first, $A = aG$, $C = cG$, and a and c are prime to each other.

Then, G is the GCD of A and C.

Suppose next, $G = gD$, $B = bD$, and g and b are prime to each other.

Then, D is the GCD of G and B, so we get: $(A$ **&** $C)$ **&** $B = D$.

Meanwhile, $A = aG$, and $C = cG$, so we get: $A = aG = agD$, and $C = cgD$ since $G = gD$.

Thus, we get: $A = agD$, $C = cgD$, and $B = bD$.

So we get: A **&** $C = gD$, which is G, of course, since A **&** C is the GCD of A and C.

Thus, we get: $(A$ **&** $C)$ **&** $B = gD$ **&** $bD = D$ since D is the GCD of gD **&** bD.

Now, what is the GCD of $A = agD$, $C = cgD$, and $B = bD$?

It is D, of course. So $(A$ **&** $C)$ **&** B is nothing but the GCD of A, B, and C.

Now, we have:

$A = x^4 + 2x^3 - x - 2$ factorized to $(x - 1)(x + 2)(x^2 + x + 1)$.

$B = 2x^4 + 3x^3 - x^2 - 4$ factorized to $(x - 1)(x + 2)(2x^2 + x + 2)$.

$C = x^3 + 7x^2 + 11x + 2$ factorized to $(x + 2)(x^2 + 5x + 1)$.

So the GCD of A, B, and C is $x + 2$.

In fact, the same is true for $(A \& B) \& C$ and $(B \& C) \& A$, too.   How come?

Suppose first, $A = aG$, $B = bG$, and a and b are prime to each other.

Then, G is the GCD of A and B. That is, $A \& B = G$.

Suppose next, $G = gD$, $C = cD$, and g and c are prime to each other.

Then, D is the GCD of G and C, so we get: $(A \& B) \& C = D$.

Meanwhile, $A = aG$, and $B = bG$, so we get: $A = aG = agD$, and $B = bgD$ since $G = gD$.

Thus, we get: $A = agD$, $B = bgD$, and $C = cD$. So D is the GCD of A, B, and C.

Thus, $(A \& B) \& C$ is the GCD of A, B, and C, too.

Suppose next, $B = bG$, $C = cG$, and b and c are prime to each other.

Then, G is the GCD of B and C. That is, $B \& C = G$.

Suppose next, $G = gD$, $A = aD$, and g and a are prime to each other.

Then, D is the GCD of G and A, so we get: $(B \& C) \& A = D$.

Meanwhile, $B = bG$, and $C = cG$, so we get: $B = bG = bgD$, and $C = cgD$ since $G = gD$.

Thus, we get: $A = aD$, $B = bgD$, and $C = cgD$. So D is the GCD of A, B, and C.

Thus, **(B & C) & A** is the GCD of *A*, *B*, and *C*, too.

Therefore, **(A & B) & C = (B & C) & A = (C & A) & B** is the GCD of *A*, *B*, and *C*.

  • Let's next, move on to **(B & C) & B**.

Suppose first, **B = bG**, **C = cG**, and *b* and *c* are prime to each other.

Then, *G* is the GCD of *B* and *C*. That is, **B & C = G**.

Thus, we get: **(B & C) & B = G & B = G** since **B = bG**, and **B & C = G**, so **G & bG = G**.

In fact, we get **(B & C) & C = G**, too, since **C = cG**, and **B & C = G**, so **G & cG = G**.

So this time, too, we want to actually find **G**, which is the GCD of **B** and **C**.

Now, we have: $A = (x - 1)(x + 2)(x^2 + x + 1)$, $B = (x - 1)(x + 2)(2x^2 + x + 2)$, and $C = (x + 2)(x^2 + 5x + 1)$. Thus, the GCD of **B** and **C** is $x + 2$.   So **(B & C) & B** $= x + 2$.

In short:

Let's first, factorize all the three polynomials, which are as follows:

$A = x^4 + 2x^3 - x - 2$, $B = 2x^4 + 3x^3 - x^2 - 4$, and $C = x^3 + 7x^2 + 11x + 2$.

First, evaluating $A = x^4 + 2x^3 - x - 2$ for $x = 1$, we get:
$1^4 + 2 \cdot 1^3 - 1 - 2 = 1 + 2 - 1 - 2 = 0$.

So $x - 1$ is a divisor of *A*. Thus, doing the synthetic division, we get:

1	1	2	0	-1	-2
		1	3	3	2
	1	3	3	2	0

So we get: $A = x^4 + 2x^3 - x - 2 = (x - 1)(x^3 + 3x^2 + 3x + 2)$.

Next, evaluating $x^3 + 3x^2 + 3x + 2$ for $x = -2$, we get:

$(-2)^3 + 3 \cdot (-2)^2 + 3(-2) + 2 = -8 + 12 - 6 + 2 = 0.$

So $x + 2$ is a divisor of $x^3 + 3x^2 + 3x + 2$. Thus, doing the synthetic division, we get:

-2	1	3	3	2
		-2	-2	-2
	1	1	1	0

So we get: $x^3 + 3x^2 + 3x + 2 = (x + 2)(x^2 + x + 1).$

Thus, $A = x^4 + 2x^3 - x - 2 = (x - 1)(x^3 + 3x^2 + 3x + 2) = (x - 1)(x + 2)(x^2 + x + 1).$

Next, evaluating $B = 2x^4 + 3x^3 - x^2 - 4$ for $x = 1$, we get:

$2 \cdot 1^4 + 3 \cdot 1^3 - 1 - 4 = 5 - 5 = 0.$

So $x - 1$ is a divisor of $2x^4 + 3x^3 - x^2 - 4$. Thus, doing the synthetic division, we get:

1	2	3	-1	0	-4
		2	5	4	4
	2	5	4	4	0

So we get: $B = 2x^4 + 3x^3 - x^2 - 4 = (x - 1)(2x^3 + 5x^2 + 4x + 4).$

Next, evaluating $2x^3 + 5x^2 + 4x + 4$ for $x = -2$, we get:

$2(-2)^3 + 5 \cdot (-2)^2 + 4(-2) + 4 = -16 + 20 - 8 + 4 = 0.$

So $x + 2$ is a divisor of $2x^3 + 5x^2 + 4x + 4$. Thus, doing the synthetic division, we get:

-2	2	5	4	4
		-4	-2	-4
	2	1	2	0

So we get: $2x^3 + 5x^2 + 4x + 4 = (x + 2)(2x^2 + x + 2).$

Thus, $B = 2x^4 + 3x^3 - x^2 - 4 = (x - 1)(2x^3 + 5x^2 + 4x + 4) = (x - 1)(x + 2)(2x^2 + x + 2).$

Next, evaluating $x^3 + 7x^2 + 11x + 2$ for $x = -2$, we get:

$(-2)^3 + 7(-2)^2 + 11(-2) + 2 = -8 + 28 - 22 + 2 = 0$.

So $x + 2$ is a divisor of $x^3 + 7x^2 + 11x + 2$. Thus, doing the synthetic division, we get:

-2	1	7	11	2
		-2	-10	-2
	1	5	1	0

Thus, dividing $x^3 + 7x^2 + 11x + 2$ by $x + 2$, we get: $x^2 + 5x + 1$ as the quotient.

Therefore, $C = x^3 + 7x^2 + 11x + 2 = (x + 2)(x^2 + 5x + 1)$.

So we now have:

$A = (x - 1)(x + 2)(x^2 + x + 1)$.

$B = (x - 1)(x + 2)(2x^2 + x + 2)$.

$C = (x + 2)(x^2 + 5x + 1)$.

So the GCD of A, B, and C is: $x + 2$.

Let's now, find $(A \& B) \& A$, $(A \& C) \& B$, $(B \& C) \& A$, and $(B \& C) \& B$.

Suppose $A = aG$, $B = bG$, and a and b are prime to each other.

Then, $(A \& B) \& A = G \& A = G$, which is the GCD of A and B, which is $(x - 1)(x + 2)$.

Let's next, move on to $(A \& C) \& B$.

Suppose first, $A = aG$, $C = cG$, and a and c are prime to each other.

Then, G is the GCD of A and C.

Suppose next, $G = gD$, $B = bD$, and g and b are prime to each other.

Then, D is the GCD of G and B, so we get: $(A$ & $C)$ & $B = D$.

Meanwhile, $A = aG$, and $C = cG$, so we get: $A = aG = agD$, and $C = cgD$ since $G = gD$.

Thus, we get: $A = agD$, $C = cgD$, and $B = bD$.

So we get: A & $C = gD$, which is G.

Thus, we get:

$(A$ & $C)$ & $B = gD$ & $bD = D$, which is the GCD of $A = agD$, $C = cgD$, and $B = bD$.

The GCD of A, B, and C is $x + 2$.

And in fact, the same is true for $(B$ & $C)$ & A, too.   The reason is as follows:

Suppose $B = bG$, $C = cG$, and b and c are prime to each other.
Then, G is the GCD of B and C. That is, B & $C = G$.

Suppose next, $G = gD$, $A = aD$, and g and a are prime to each other.
Then, D is the GCD of G and A, so we get: $(B$ & $C)$ & $A = D$.

Meanwhile, $B = bG$, and $C = cG$, so we get: $B = bG = bgD$, and $C = cgD$ since $G = gD$.

Thus, we get: $A = aD$, $B = bgD$, and $C = cgD$.

So D is the GCD of A, B, and C.

Thus, $(B$ & $C)$ & A is the GCD of A, B, and C, too.

Let's next, move on to (**B & C**) **& B**.

Suppose **B** = **bG**, **C** = **cG**, and **b** and **c** are prime to each other.

Then, **G** is the GCD of **B** and **C**.

Thus, (**B & C**) **& B** = **G & B** = **G**, which is the GCD of **B** and **C**, which is $x + 2$.

Examples 6 in GCD and LCM

0.   Assuming $A = x^3 - 2x^2 - 5x + 6$, and $B = x^3 + a^2x^2 + 2ax - 16$, find all the values of a in each of the cases below:

0.0.   The degree of the GCD of A and B is ≥ 1.

0.1.   The degree of the GCD of A and B is 2.

1.   Find the GCD of 69300 and 382200.

Suggestions or Solutions
To the **Problem 0** in the Example **0**

Assuming $A = x^3 - 2x^2 - 5x + 6$, and $B = x^3 + a^2x^2 + 2ax - 16$, **find all the values of a in the case where the degree of the GCD of A and B is ≥ 1.**

First, $x = 1 \Rightarrow A = x^3 - 2x^2 - 5x + 6 = 1^4 + 2 \cdot 1^3 - 1 - 2 = 1 + 2 - 1 - 2 = 0$.

So $x - 1$ is a divisor of A. Thus, doing the synthetic division, we get:

$$
\begin{array}{r|rrrr}
1 & 1 & -2 & -5 & 6 \\
  &   & 1 & -1 & -6 \\
\hline
  & 1 & -1 & -6 & 0
\end{array}
$$

So we get: $A = x^3 - 2x^2 - 5x + 6 = (x - 1)(x^2 - x - 6) = (x - 1)(x + 2)(x - 3)$.

Thus, either of $x - 1$, $x + 2$, and $x - 3$ has to be a factor common to A and B.

Assuming first, $x - 1$ is a divisor common to A and B, we get:

$x = 1 \Rightarrow B = 1 + a^2 + 2a - 16 = a^2 + 2a - 15 = (a + 5)(a - 3) = 0 \Rightarrow a = $ -5 or 3.

Assuming next, $x + 2$ is a divisor common to A and B, we get:

$x = $ -2 $\Rightarrow B = $ -8 $+ 4a^2 - 4a - 16 = 4a^2 - 4a - 24 = 4(a - 3)(a + 2) = 0 \Rightarrow a = 3$ or -2.

And assuming next, $x - 3$ is a divisor common to A and B, we get:

$x = 3 \Rightarrow 27 + 9a^2 + 6a - 16 = 9a^2 + 6a + 11 = (3a + 1) + 10 \geq 10$.

So $x - 3$ cannot be a divisor of B.

Therefore, $a = 3$, **-2**, or **-5**.

If not quite sure of the idea behind the processes above, follow the steps below:

To begin with, what do we mean by the degree of a polynomial?

It is the largest exponent applied to the variable the polynomial is subject to.

So A can be called a polynomial of degree 5.
And a monomial can have a degree, too. For instance, $3x^2$ is a monomial of degree 2.
What then, about the degree of the GCD?

The GCD of polynomials can be an integer (or constant), a monomial, or a polynomial.

And in particular, an integer (or a constant) can be said to be of degree 0.

That's because we have: $5x^0 = 5$, since $x^0 = 1$.
And assuming a is constant, we get: $ax^0 = a$, and $(a + 1)x^0 = a + 1$, which is constant, since a is constant. So if the degree of the GCD is 0, the GCD is an integer (or constant).

And if for instance, the degree of the GCD is 3, the GCD is a monomial of degree 3 as $2x^3$ or a polynomial of degree 3 as $5x^3 + x - 1$.

Now, this problem is talking about GCD, which is about divisors, that is, factors.

So let's first, factorize the polynomial $A = x^3 - 2x^2 - 5x + 6$.

First, evaluating $A = x^3 - 2x^2 - 5x + 6$ for $x = 1$, we get:
$1^4 + 2 \cdot 1^3 - 1 - 2 = 1 + 2 - 1 - 2 = 0$. How to get $x = 1$ is covered in the previous example.

So $x - 1$ is a divisor of A. Thus, doing the synthetic division, we get:

1	1	-2	-5	6
		1	-1	-6
	1	-1	-6	0

So we get: $A = x^3 - 2x^2 - 5x + 6 = (x - 1)(x^2 - x - 6)$.

Next, we want to check to see if the quotient $x^2 - x - 6$ can still be factorized.

In fact, $x^2 - x - 6$ can be factorized to $(x - 3)(x + 2)$.

Therefore, $A = x^3 - 2x^2 - 5x + 6$ gets factorized to $(x - 1)(x + 2)(x - 3)$.

Besides, knowing the fact that the divisors of A can be $x - 1$, $x + 2$, and $x - 3$ ahead of time, we can do synthetic divisions sequentially in such a way as follows:

1	1	-2	-5	6
		1	-1	-6
-2	**1**	**-1**	**-6**	0
		-2	6	
	1	-3	0	

So having only to look at the arrays of numbers above, we can see that:

- $(x - 1)$ is a divisor, and the quotient is: $x^2 - x - 6$ from **1**, **-1**, and **-6**.

- $\{x - (-2)\}$, that is, $(x + 2)$ is a divisor, too, and the quotient is: $x - 3$ from 1 and -3.

- Now, in this problem, we have: $A = (x - 1)(x + 2)(x - 3)$, and $B = x^3 + a^2x^2 + 2ax - 16$.

And we have the fact that the degree of the GCD of A and B is ≥ 1.

That is, the degree of the GCD is at least 1.   What then, can the GCD be?

The GCD is an expression monomial or polynomial, and the degree of the expression is at least 1. So we may want to see first, what the expression is.

To begin with, the GCD of A and B is the greatest of all divisors common to A and B.

That is, the greatest divisor common to A and B is the GCD of A and B.

And we know that a factor is a divisor.
So if the GCD exists, a factor common to A and B has to exist anyway.
Thus, at least one of the factors of A has to be the factor of the polynomial B.   So?

So we want to look at the factors of A.

And we have: $A = (x - 1)(x + 2)(x - 3)$.

So displaying all the possible factors of A, we get:

$(x - 1)(x + 2)(x - 3)$, $(x - 1)(x + 2)$, $(x - 1)(x - 3)$, $(x + 2)(x - 3)$, $x - 1$, $x + 2$, and $x - 3$.

Then, the GCD is a polynomial, which is one of the seven factors above.
And we know that the GCD is an expression of degree at least 1.

So of the seven factors above, either of the last three has to be common to A and B.

And the three are: $x - 1$, $x + 2$, and $x - 3$.
Thus, at least one of the three above is a common factor of A and B.

So let's begin with $x - 1$.
Suppose now, $x - 1$ is a common factor, that is, a divisor common to A and B.

What then, is B when $x = 1$?

It has to be 0, since $x - 1$ is a divisor of B.
That is, B has to be in a form of $(x - 1)Q$, where Q is of degree 2, for B is of degree 3.
So?

We have: $B = x^3 + a^2x^2 + 2ax - 16$.
So we get: $x = 1 \Rightarrow B = 1 + a^2 + 2a - 16 = a^2 + 2a - 15 = (a + 5)(a - 3) = 0$.

Therefore, we get: $a = -5$ or 3.

Suppose next, $x + 2$ is a common factor, that is, a divisor common to A and B. Then, what is B when $x = -2$?

It has to be 0, too, since $x - 2$ is a divisor of B.

That is, B has to be put in $(x + 2)Q$, where Q is of degree 2, since B is of degree 3.

And we have $B = x^3 + a^2x^2 + 2ax - 16$.

So we get: $x = -2 \Rightarrow B = -8 + 4a^2 - 4a - 16 = 4a^2 - 4a - 24 = 4(a - 3)(a + 2) = 0$.

Therefore, we get: $a = 3$ or -2.

And suppose next, $x - 3$ is a factor common to A and B.   What then, is B when $x = 3$?

It has to be 0, too, of course, since $x - 3$ is a divisor of B.

That is, B needs to be put in $(x - 3)Q$, where Q is of degree 2, since B is of degree 3.

And since $B = x^3 + a^2x^2 + 2ax - 16$, we get:

$x = 3 \Rightarrow B = 27 + 9a^2 + 6a - 16 = 9a^2 + 6a + 11 = (3a + 1)^2 + 10 \geq 10$.   What then?

This time, B cannot be 0 for any value of a, and thus, $x - 3$ cannot be a divisor of B.

And thus, putting threads together, we get: $a = 3$, -2, or -5.

Suggestions or Solutions
To the **Problem 1** in the Example **0**

Assuming $A = x^3 - 2x^2 - 5x + 6$, and $B = x^3 + a^2x^2 + 2ax - 16$, find all the values of a in the case where the degree of the GCD of A and B is 2.

If $a = 3$ or -5, $x - 1$ is a factor common to A and B.

If $a = 3$ or -2, $x + 2$ is a common factor.

So $(x - 1)(x + 2)$ is the GCD, and satisfying both conditions above, $a = 3$ only.

If not quite sure of the idea behind the processes above, follow the steps below:

In this problem, we have: $A = (x - 1)(x + 2)(x - 3)$, and $B = x^3 + a^2x^2 + 2ax - 16$, along with the fact that the degree of the GCD of A and B is 2.

So the GCD is a polynomial of degree 2, which is a factor common to A and B.
Thus, we may want to get back to the factors of A, and they are as follows:

$(x - 1)(x + 2)(x - 3)$, $(x - 1)(x + 2)$, $(x - 1)(x - 3)$, $(x + 2)(x - 3)$, $x - 1$, $x + 2$, and $x - 3$.

Of the seven factors above, either of the three below has to be common to A and B:

$(x - 1)(x + 2)$, $(x - 1)(x - 3)$, and $(x + 2)(x - 3)$.

That's because the degree of the GCD is 2.
Thus, at least one of the three above is a common factor of A and B.

Of the three however, the last two cannot be applicable.
How come?

We have facts as follows:

If $a = 3$ or -5, then $x - 1$ can be a factor common to A and B.

If $a = 3$ or -2, then $x + 2$ can be a common factor.

However, $x - 3$ cannot be a common factor for any value of a.   So?

If $(x - 1)(x - 3)$ is a common factor, $(x - 1)$ and $(x - 3)$ both have to be common factors.

If $(x + 2)(x - 3)$ is a common factor, $(x + 1)$ and $(x - 3)$ both have to be common factors.

So neither of $(x - 1)(x - 3)$ and $(x + 2)(x - 3)$ can be applicable.

What then, about $(x - 1)(x + 2)$?

It is in fact, the GCD, because both of $(x - 1)$ and $(x + 2)$ can be common factors.

What then, about the value of a?

We know: $(x - 1)$ and $(x + 2)$ can be common factors, and both have to be common factors at the same time if $(x - 1)(x + 2)$ is a common factor.

So the value of a has to satisfy both conditions that $x - 1$ and $x + 2$ both are factors common to A and B at the same time. And we have the fact that:

If $x - 1$ is a factor common to A and B, we get: $a = 3$ or -5.

If $x + 2$ is a common factor, we get: $a = 3$ or -2.   So?

So satisfying both cases above, we need to have: $a = 3$ only.

That is, only if $a = 3$, $(x - 1)(x + 2)$ is the GCD.

Suggestions or Solutions
To the **Problem** in the Example 1

Find the GCD of 69300 and 382200.

Usually, finding the GCD of a set of integers, we take a product of all powers, in each of which the base is a prime factor common to all the integers, and the exponent is the smallest of all used for the prime factor.

So the product has all prime factors common to all the integers, and thus, is the divisor the greatest and common to all the integers, that is, the GCD. And the same is true for monomials and polynomials, too.

So what should we begin with finding the GCD of the two integers given?

We want to begin with factorizing the two integers.

So factorizing both integers given, we get: $69300 = 2^2 3^2 5^2 7 \cdot 11$, and $382200 = 2^3 3 \cdot 5^2 7^2 13$.

• Next, finding all the prime factors common to both integers, we get: 2, 3, 5, and 7.

• Next, finding the smallest of all exponents used for each of the prime factors above, we get: 2 for the prime factor 2, 1 for the factor 3, 2 for 5, and 1 for 7.

• Next, taking the product of all the powers, we get: $2^2 3 \cdot 5^2 7$, which is the GCD, which is 2100 if expanded.

Besides, we can use a theorem regarding the GCD, and the theorem is as follows:

• Suppose dividing A by B, we get Q as the quotient, and R as the remainder.

Then, we get: $A = BQ + R$, and the GCD of A and B = the GCD of B and R.

That is, the GCD of a dividend and a divisor is the GCD of the divisor and the remainder.

So if R divides B, R is the GCD of A and B.   How come?

The theorem has been proved in the Example 3 earlier.

Let's however, see now again, how the theorem works.

Suppose A and B are integers, $A = aG$, $B = bG$, and G is the GCD of A and B.

Then, a and b are prime to each other, and we can get:

$A = BQ + R \Rightarrow aG = bGQ + R \Rightarrow R = (a - bQ)G$.

Thus, we now have: $B = bG$, and $R = (a - bQ)G$, where G is the GCD of B and R.

So b and $(a - bQ)$ are prime to each other, since G is the GCD of B and R.

Suppose now, $B = kR$, where k is an integer.
That is, R is a factor of B. In other words, R divides B.

Then, $B = kR \Rightarrow k = \frac{B}{R} = \frac{bG}{(a-bQ)G} = \frac{b}{a-bQ}$, so $(a - bQ)$ has to be 1.   How come?

That's because if $a - bQ \neq 1$, k cannot be an integer, because b and $(a - bQ)$ are prime to each other. So we need to have: $a - bQ = 1$ if k is an integer.

Also, even though $a - bQ = 1$, b and $(a - bQ)$ are still prime to each other, because 1 is the only divisor common to b and $(a - bQ)$ which is 1.

So in the case where $B = bG$, and $R = (a - bQ)G$, if we get: $B = kR$ where k is an integer, we get: $a - bQ = 1$, so we get: $R = G$, which is the GCD.

And the same is true, too, for cases where A and B are monomials or polynomials.

So using the theorem above, we can get the GCD the way below, too:

First, 382200 = 69300·5 + 35700, where 69300 is **B** in the theorem, and 35700 is **R**.

So next, 69300 = 35700·1 + 33600, where 35700 is **B**, and 33600 is **R**.

So next, 35700 = 33600·1 + 2100, where 33600 is **B**, and 2100 is **R**.

So next, 33600 = 2100·16 + 0, and thus, **R** above divides **B** above.

Thus, **R**, which is 2100 is the GCD. And let's see now, if it really is the case:

We get: $33600 = 21·100·16 = 3·7·2^2·5^2 2^4 = 2^4\underline{2^2 3·5^2 7}$, and $2100 = 3·7·2^2·5^2 = \underline{2^2 3·5^2 7}$.

So the GCD of 33600 and 2100 is $2^2 3·5^2 7$, which is the GCD of the original two integers 382200 and 69300.

And let's for reference, put the processes above the way below:

First, 382200 = 69300·5 + 35700, where 5 is the quotient, and 35700 is the remainder. That is, $2^3 3·5^2 7^2 13 = (2^2 3^2 5^2 7·11)·5 + 2^2 5^2 3·7·17$.

So next, 69300 = 35700·1 + 33600, where 1 is the quotient, and 33600 is the remainder. That is, $2^2 3^2 5^2 7·11 = (2^2 5^2 3·7·17)·1 + 2^6 5^2 3·7$.

So next, 35700 = 33600·1 + 2100, where 1 is the quotient, and 2100 is the remainder. That is, $2^2 5^2 3·7·17 = (2^6 5^2 3·7)·1 + 2^2 5^2 3·7$.

So next, 33600 = 2100·16 + 0, where 16 is the quotient, and 0 is the remainder. That is, $2^6 5^2 3·7 = (2^2 5^2 3·7)·2^4 + 0$.

So $2^2 5^2 3·7$ is the GCD, which is the GCD of $382200 = 2^3 3·5^2 7^2 13$, and $2100 = 2^2 3·5^2 7$.

Examples 7 in GCD and LCM

0. Find the GCD of $x^4 + 3x^3 - 2x^2 - 3x - 5$ and $x^3 + 7x^2 + 7x + 6$.

1. Suppose $x^2 + 3x + 2$ is the GCD of two polynomials, $x^4 + 5x^3 - 7x^2 - 41x - 30$ is the LCM of the two, and 1 is the coefficient of the highest term in each of the two.

Then, find the two polynomials.

Suggestions or Solutions
To the **Problem** in the Example **0**

Find the GCD of $x^4 + 3x^3 - 2x^2 - 3x - 5$ **and** $x^3 + 7x^2 + 7x + 6$.

Suppose $P = x^4 + 3x^3 - 2x^2 - 3x - 5$, and $B = x^3 + 7x^2 + 7x + 6$.

Then first, we can set $P = BQ + R$, where Q is a quotient, and R is the remainder.

So we can do a division as below:

$$
\begin{array}{r}
x - 4 \\
\hline
x^3 + 7x^2 + 7x + 6 \enclose{longdiv}{x^4 + 3x^3 - 2x^2 - 3x - 5} \\
x^4 + 7x^3 + 7x^2 + 6x \\
\hline
-4x^3 - 9x^2 - 9x - 5 \\
-4x^3 - 28x^2 - 28x - 24 \\
\hline
19x^2 + 19x + 19
\end{array}
$$

Thus, $P = B(x - 4) + 19x^2 + 19x + 19$.

So setting $r = x^2 + x + 1$, we get: $P = B(x - 4) + 19r \Rightarrow P = B(x - 4) + 19r$.

Next, we can set $B = rq + R$, where q is a quotient, and R is the remainder.

So we can do a division as below:

$$
\begin{array}{r}
x + 6 \\
\hline
x^2 + x + 1 \enclose{longdiv}{x^3 + 7x^2 + 7x + 6} \\
x^3 + x^2 + x \\
\hline
6x^2 + 6x + 6 \\
6x^2 + 6x + 6 \\
\hline
0
\end{array}
$$

Thus, $B = x^3 + 7x^2 + 7x + 6 = (x^2 + x + 1)(x + 6) = r(x + 6) \Rightarrow B = r(x + 6)$.

So we get: $P = B(x - 4) + 19r = r(x + 6)(x - 4) + 19r = r\{(x + 6)(x - 4) + 19\}$
$= r(x^2 + 2x - 24 + 19) = r(x^2 + 2x - 5) \Rightarrow P = r(x^2 + 2x - 5)$.

Therefore, the GCD is r, which is $x^2 + x + 1$.

If not quite sure of the idea behind the processes above, follow the steps below:

Usually, finding the GCD of a set of polynomials, we take a product of all powers, in each of which, the base is a prime factor common to all the polynomials, and the exponent is the smallest of all used for the prime factor.

So the product has all the prime factors common to all the polynomials, and thus, is the divisor the greatest and common to all the polynomials, that is, the GCD.

Therefore, we want to begin with factorizing all the polynomials.
Factorizing though, such a polynomial as $x^4 + 3x^3 - 2x^2 - 3x - 5$ is not easy.
In fact, no divisor of 5 can be the value of x that can make the polynomial 0. So in this case, we cannot find an integer d for which $x - d$ can divide the polynomial above.

When we solve this kind of problem, too, though, the theorem used in the example 1 in **Examples 6** can be quite handy, and is as follows:

Suppose dividing A by B, we get Q as the quotient, and R as the remainder.
Then, we get: $A = BQ + R$, and the GCD of A and B is the GCD of B and R.
That is, the GCD of a dividend and a divisor is the GCD of the divisor and the remainder.
So if R divides B, R is the GCD of A and B.

And we have seen how it is the case in the example 1 in **Examples 6**.

Let's however, see now again, how R is the GCD of A and B if R divides B.

Note first, that normally, factorizing polynomials, we assume that factors are integers or expressions with integer coefficients. So for instance, we consider as a factor $2x + y + 1$, $3y^2$, or $2c$ and not $\frac{3}{2}x + y + 1$, $\frac{1}{3}y^2$, or $\frac{2c}{5}$.

Now, suppose, A and B are polynomials, $A = aG$, $B = bG$, and G is the GCD of A and B. Then, a and b are prime to each other, and we can get:
$A = BQ + R \Rightarrow aG = bGQ + R \Rightarrow R = (a - bQ)G$.

So we now have: $B = bG$, and $R = (a - bQ)G$, where G is the GCD of B and R.

Therefore, b and $(a - bQ)$ both can be integers, constants, monomials, or polynomials, and are prime to each other, since G is the GCD of B and R.

Suppose now, $B = kR$, where k is an integer, constant, monomial, or polynomial. That is, k and R are factors of B. In other words, each of k and R divides B.

Then, we get: $B = kR \Rightarrow k = \frac{B}{R} = \frac{bG}{(a-bQ)G} = \frac{b}{a-bQ}$, so $(a - bQ)$ has to be 1.   How come?

If $a - bQ \neq 1$, k cannot be a factor positive, constant, monomial, or polynomial, because b and $(a - bQ)$ are integers (or constants), monomials, or polynomials, and are prime to each other.   How come?

First, b is an integer (or constant), monomial, or polynomial, and so is $(a - bQ)$.

So next, if $a - bQ \neq 1$, since b and $(a - bQ)$ are prime to each other, $k = \frac{b}{a-bQ}$ cannot be a factor that is an integer, a constant, a monomial, or a polynomial, but $k = \frac{b}{a-bQ}$ is a fractional number as $\frac{3}{2}$ or expression as $\frac{2x}{x-y}$, $\frac{3}{2x-y}$, or $\frac{2c}{y}$, which is neither a monomial nor a polynomial. Technically of course, $\frac{3c}{2}$, for instance, can be taken as a constant, too, if c is a constant, but is a fractional constant.

Thus, $(a - bQ)$ has to be 1 if k is a factor that is integer (or constant), monomial, or polynomial.

Also, even though $a - bQ = 1$, b and $(a - bQ)$ are still prime to each other, because 1 is the only divisor common to b and $(a - bQ)$ which is 1.

So in the case where $B = bG$, and $R = (a - bQ)G$, if we get: $B = kR$ where k is an integer, a constant, a monomial, or a polynomial, we get: $a - bQ = 1$, so we get: $R = G$, which is the GCD.

So using the theorem above, we can get the GCD in such a way as follows, too:

Suppose $P = x^4 + 3x^3 - 2x^2 - 3x - 5$, and $B = x^3 + 7x^2 + 7x + 6$.

Then first, we can set: $P = BQ + R$, where Q is a quotient, and R is the remainder.

So we can do a division as below:

$$
\begin{array}{r}
x - 4 \\ \hline
x^3 + 7x^2 + 7x + 6 \overline{\smash{\big)}\ x^4 + 3x^3 - 2x^2 - 3x - 5} \\
\underline{x^4 + 7x^3 + 7x^2 + 6x} \\
-4x^3 - 9x^2 - 9x - 5 \\
\underline{-4x^3 - 28x^2 - 28x - 24} \\
19x^2 + 19x + 19
\end{array}
$$

Thus, $P = (x^3 + 7x^2 + 7x + 6)(x - 4) + 19x^2 + 19x + 19$, where $x^3 + 7x^2 + 7x + 6$ is B in the theorem, and $19x^2 + 19x + 19$ is R.

So next, $x^3 + 7x^2 + 7x + 6 = (19x^2 + 19x + 19)Q + R$, where $19x^2 + 19x + 19$ is now B in the theorem, and R is of course, R in the theorem.

Meanwhile though, we have: $19x^2 + 19x + 19 = 19(x^2 + x + 1)$.

So setting $r = x^2 + x + 1$, we get: $x^3 + 7x^2 + 7x + 6 = 19rQ + R = r19Q + R$.

Now, anyway, Q above simply indicates just a quotient.

Thus, setting: $q = 19Q$, we get: $x^3 + 7x^2 + 7x + 6 = (x^2 + x + 1)q + R$, where q merely indicates a quotient, too.

So $x^2 + x + 1$ is now B in the theorem, and R is of course, R in the theorem.

And thus, doing the division, we get:

$$\begin{array}{r} x + 6 \\ x^2 + x + 1 \overline{\smash{)}\ x^3 + 7x^2 + 7x + 6} \\ \underline{x^3 + x^2 + x} \\ 6x^2 + 6x + 6 \\ \underline{6x^2 + 6x + 6} \\ 0 \end{array}$$

Thus, we get: $x^3 + 7x^2 + 7x + 6 = (x^2 + x + 1)(x + 6) + 0$, so the quotient is $x + 6$, and the remainder is 0. So $x^2 + x + 1$ which is R in the theorem, divides $x^3 + 7x^2 + 7x + 6$ which is B in the theorem. Thus, we can see that $x^2 + x + 1$ is the GCD.

And let's now, put the processes above the way below:

To begin with, we set: $P = x^4 + 3x^3 - 2x^2 - 3x - 5$, and $B = x^3 + 7x^2 + 7x + 6$.

In the first step, dividing P by B, we get: $P = B(x - 4) + 19x^2 + 19x + 19$.

Next, we set: $B = x^3 + 7x^2 + 7x + 6 = (19x^2 + 19x + 19)Q + R$.

Meanwhile, we set: $19x^2 + 19x + 19 = 19r$, and $q = 19Q$. So we get: $B = rq + R$.

In the second step, dividing B by r, we get: $B = x^3 + 7x^2 + 7x + 6 = (x^2 + x + 1)(x + 6)$.

So we can notice that $x^2 + x + 1$ is the GCD of P and B since the remainder is 0, that is, $x^2 + x + 1$ divides $x^3 + 7x^2 + 7x + 6$.

Now, we have:

$P = B(x - 4) + 19x^2 + 19x + 19 = B(x - 4) + 19r \Rightarrow P = B(x - 4) + 19r$.
$B = x^3 + 7x^2 + 7x + 6 = (x^2 + x + 1)(x + 6) = r(x + 6) \Rightarrow B = r(x + 6)$.

So putting threads together, we get:

$P = B(x - 4) + 19r = r(x + 6)(x - 4) + 19r = r\{(x + 6)(x - 4) + 19\} = r(x^2 + 2x - 24 + 19)$
$= (x^2 + x + 1)(x^2 + 2x - 5)$.

Thus, we can see that $x^4 + 3x^3 - 2x^2 - 3x - 5$ gets factorized to $(x^2 + x + 1)(x^2 + 2x - 5)$.

And of course, $B = x^3 + 7x^2 + 7x + 6$ gets factorized to $(x^2 + x + 1)(x + 6)$.

Therefore, the GCD is: $x^2 + x + 1$.

In short:

Suppose $P = x^4 + 3x^3 - 2x^2 - 3x - 5$, and $B = x^3 + 7x^2 + 7x + 6$.

Then first, we can set $P = BQ + R$, where Q is a quotient, and R is the remainder.

So we can do a division as below:

$$
\begin{array}{r}
x - 4 \\
\hline
x^3 + 7x^2 + 7x + 6\,\big)\, x^4 + 3x^3 - 2x^2 - 3x - 5 \\
x^4 + 7x^3 + 7x^2 + 6x \\
\hline
-4x^3 - 9x^2 - 9x - 5 \\
-4x^3 - 28x^2 - 28x - 24 \\
\hline
19x^2 + 19x + 19
\end{array}
$$

Thus, $P = B(x - 4) + 19x^2 + 19x + 19$.

So setting $r = x^2 + x + 1$, we get: $P = B(x - 4) + 19r \Rightarrow P = B(x - 4) + 19r$.

Next, we can set $B = rq + R$, where q is a quotient, and R is the remainder.

So we can do a division as below:

$$
\begin{array}{r}
x + 6 \\
\hline
x^2 + x + 1\,\big)\, x^3 + 7x^2 + 7x + 6 \\
x^3 + x^2 + x \\
\hline
6x^2 + 6x + 6 \\
6x^2 + 6x + 6 \\
\hline
0
\end{array}
$$

Thus, $B = x^3 + 7x^2 + 7x + 6 = (x^2 + x + 1)(x + 6) = r(x + 6) \Rightarrow B = r(x + 6)$.

So we get: $P = B(x - 4) + 19r = r(x + 6)(x - 4) + 19r = r\{(x + 6)(x - 4) + 19\}$

$= r(x^2 + 2x - 24 + 19) = r(x^2 + 2x - 5) \Rightarrow P = r(x^2 + 2x - 5)$.

Therefore, the GCD is r, which is $x^2 + x + 1$.

Suggestions or Solutions
To the **Problem** in the Example **1**

Suppose $x^2 + 3x + 2$ is the GCD of two polynomials, $x^4 + 5x^3 - 7x^2 - 41x - 30$ is the LCM of the two, and 1 is the coefficient of the highest term in each of the two. Then, find the two polynomials.

Suppose $G = x^2 + 3x + 2$, $L = x^4 + 5x^3 - 7x^2 - 41x - 30$, the two polynomials are A and B, and a and b are prime to each other.

Then, we get: $A = aG$, $B = bG$, and $L = abG$.

$$
\begin{array}{r}
x^2 - 2x - 15 \\
x^2 + 3x + 2 \overline{\smash{\big)}\ x^4 + 5x^3 - 7x^2 - 41x - 30} \\
\underline{x^4 + 3x^3 + 2x^2} \\
2x^3 - 9x^2 - 41x - 30 \\
\underline{2x^3 - 6x^2 + 4x} \\
-15x^2 - 45x + 30 \\
\underline{-15x^2 - 45x + 30} \\
0
\end{array}
$$

Therefore, $ab = x^2 - 2x - 15$, which is factorized to $(x - 3)(x + 5)$.

Next, a and b are prime to each other, and thus, are $x - 3$ and $x + 5$, or are $(x - 3)(x + 5)$, and 1.

Thus, since $A = aG$, and $B = bG$, A and B are $x^2 + 3x + 2$ and $(x - 3)(x + 5)(x^2 + 3x + 2)$, or are $(x - 3)(x^2 + 3x + 2)$ and $(x + 5)(x^2 + 3x + 2)$.

If not quite sure of the idea behind the processes above, follow the steps below:

To begin with, what do we mean by the highest term?

Normally, it is a term using the largest of all exponents used for all the terms.

So for instance, a polynomial $x^3 + 3x^2 + x + 2$ has four terms, which are x^3, $3x^2$, x, and 2, and the term x^3 is the highest term, for the exponent 3 is the largest of all the exponents.

Thus, the degree of the polynomial indicates the degree of the highest term.

Technically though, a term itself can be a polynomial, too, as $x + (2x+1)^2 + \frac{2y+1}{3}$. In such a case, we expand all the terms in polynomial, and then, determine the highest term.

So expanding them all, we get $x + 4x^2 + 4x + \frac{2}{3}y + \frac{4}{3}$, and the highest term is $4x^2$.

And of course, a polynomial can have many highest terms, too.

For instance, expanding $x + (2x+y)^2 + \frac{2y+1}{3}$, we get $x + 4x^2 + 4xy + y^2 + \frac{2}{3}y + \frac{1}{3}$, so the highest terms are $4x^2$, $4xy$, and y^2.

So the highest term in a polynomial can be said to be a term where the number of variable multiplied together is the largest.

Suppose now, $G = x^2 + 3x + 2$, $L = x^4 + 5x^3 - 7x^2 - 41x - 30$, the two polynomials are A and B, and a and b are prime to each other.

Then, since G is the GCD of A and B, we can set: $A = aG$, and $B = bG$, and also, we get:

$L = abG$. Thus, dividing L by G, we can find ab.

So getting ab, we can do a division as follows:

$$
\begin{array}{r}
x^2 - 2x - 15 \\
x^2 + 3x + 2 \overline{\smash{\big)}\ x^4 + 5x^3 - 7x^2 - 41x - 30} \\
\underline{x^4 + 3x^3 + 2x^2} \\
2x^3 - 9x^2 - 41x - 30 \\
\underline{2x^3 - 6x^2 + 4x} \\
-15x^2 - 45x + 30 \\
\underline{-15x^2 - 45x + 30} \\
0
\end{array}
$$

Therefore, we can see that $ab = x^2 - 2x - 15$. So we now have:

$$G = x^2 + 3x + 2, L = x^4 + 5x^3 - 7x^2 - 41x - 30, A = aG, B = bG, \text{ and } ab = x^2 - 2x - 15.$$

Next, finding a and b, we may want to factorize $x^2 - 2x - 15$.

So factorizing it, we get: $x^2 - 2x - 15 = (x - 3)(x + 5)$.

Next, we know a and b are prime to each other, so we can have two cases as follows:

Either of a and b is $x - 3$, and the other is $x + 5$.

Either of a and b is $(x - 3)(x + 5)$, and the other is 1.

Now, we have: $A = aG$, and $B = bG$.

So if first, $a = x - 3$, and $b = x + 5$, we get: $A = aG = (x - 3)G$, and $B = bG = (x + 5)G$.

If next, $a = 1$, and $b = (x - 3)(x + 5)$, we get: $A = aG = G$, and $B = bG = (x - 3)(x + 5)G$.

So the two polynomials are:

$(x - 3)G$ and $(x + 5)G$, or G and $(x - 3)(x + 5)G$.

And we have: $G = x^2 + 3x + 2$.

Thus, the two polynomials can be in either of the two cases as below:

$(x - 3)(x^2 + 3x + 2)$ and $(x + 5)(x^2 + 3x + 2)$

$x^2 + 3x + 2$ and $(x - 3)(x + 5)(x^2 + 3x + 2)$

www.ingramcontent.com/pod-product-compliance
Lightning Source LLC
Chambersburg PA
CBHW081114170526
45165CB00008B/2444